A Practical Guide to Scientific Writing in Chemistry

Successful completion of postgraduate studies, especially PhD, and career advancement in academia strongly depend on the ability to publish scientific papers or books and attract research grants. However, many chemical scientists find preparing scientific papers and research grant and book proposals difficult; partly because of insufficient training in writing and partly because there are few practical books to enable them to learn the art. This step-by-step practical guide is intended mainly for postgraduate students and early career researchers in chemical science and the libraries that serve them but will also be useful to other scientists.

Key Features:
- Improves the reader's chances of getting their manuscript published in chemistry journals.
- Increases the likelihood of winning research grants in chemistry.
- Takes a "lead by the hand" approach.
- Contains chapters on the preparation of graphical abstracts and research highlights.
- Uses sketches and other illustration styles to aid mental visualization of concepts.
- Contains practical examples taken from published papers and successful research grant proposals.

T0273338

A Practical Guide to Scientific Writing in Chemistry

Scientific Papers, Research Grants and Book Proposals

Andrew Terhemen Tyowua

Department of Chemistry, Benue State University, Makurdi, Nigeria

"No one succeeds by complaining about challenges,
All successful people succeed by surmounting challenges"

Andrew Terhemen Tyowua

CRC Press
Taylor & Francis Group
Boca Raton London New York

CRC Press is an imprint of the
Taylor & Francis Group, an **informa** business

Designed cover image: © Shutterstock

First edition published 2023
by CRC Press
6000 Broken Sound Parkway NW, Suite 300, Boca Raton, FL 33487-2742

and by CRC Press
4 Park Square, Milton Park, Abingdon, Oxon, OX14 4RN

CRC Press is an imprint of Taylor & Francis Group, LLC

© 2023 Andrew Terhemen Tyowua

ISBN: 9781032033181 (hbk)
ISBN: 9781032033204 (pbk)
ISBN: 9781003186748 (ebk)

DOI: 10.1201/9781003186748

Typeset in Times
by codeMantra

To my wife (Pharmacist Maryam), three children (Aondover, Kumawuese and Aondosoo) and my parents (Tyowua Ibier and Icher Ibier)

Contents

Preface

Scientists are always faced with the task of communicating effectively *via* writing to audiences within and outside their disciplines. As I review research grant proposals for prestigious funders and manuscripts for top scientific journals in chemistry, I have observed that writing effectively is challenging for many chemical scientists. This challenge is partly due to insufficient university training on writing and partly due to lack of practical textbooks on writing. While very little can be done about university training, the availability of practical textbooks on writing will go a long way in alleviating this challenge. Meanwhile, there are no practical textbooks on writing in chemistry, but there are numerous guides on scientific writing in the form of either journal papers or textbooks. The downside is that these guides are not practical and are biased toward other disciplines. Additionally, these guides cover mainly the writing of research papers with very little or nothing on other important scientific genres like posters, oral presentations, research grant and book proposals, meaning chemical scientists will gain very little from them. This calls for a practical textbook on writing for chemical scientists. The essence of *A Practical Guide to Scientific Writing in Chemistry* is to answer this call.

A Practical Guide to Scientific Writing in Chemistry is based on a lecture course on scientific writing and ethics, given to postgraduate students of chemistry at Benue State University, Makurdi, Nigeria. The book is divided into five sections with section one covering scientific papers (Chapters 1–19), section two covering poster presentations (Chapter 20), section three covering oral presentations (Chapter 21), section four covering research grant proposals (Chapters 22–31) and section five covering book proposals (Chapter 32). The language used throughout the book is simple and straight to the point. Equal attention is given to all the sections (title, abstracts, introduction, materials, methods, results, discussion, conclusions and references) of these genres. All examples and tasks used in the book are taken from easily understandable published materials. Sketches/illustrations are used extensively to aid mental visualization of concepts.

While effective writing cannot be reduced to a set of simple rules, writing is easier and more efficient when certain rules are followed. These rules are embodied in a clearly defined structure and choice of words. Therefore, for each genre discussed, the most appropriate structure and choice of words are emphasized. Clearly, chemical scientists will find this book more useful compared with the existing guides on scientific writing which have limited coverage and are biased toward other disciplines.

Finally, writing and giving talks are forms of art and can only be optimized by repetition like any other form of art. For example, you can read all the excellent instructional books on the art of singing, dancing, piano playing, bicycle riding, horse riding, painting, *etc.*, but unless you make repeated conscious efforts, you will not succeed at any. This also applies to writing and giving oral presentation. You can only improve by writing, editing, rewriting, presenting and presenting. With *A Practical Guide to Scientific Writing in Chemistry* by your side, these will be much easier. Although this book is written specifically for chemical scientists, other scientists can also find it useful.

Andrew Terhemen Tyowua
Benue State University, Makurdi, Nigeria

Author

Andrew Terhemen Tyowua teaches physical, inorganic and colloid chemistry and scientific writing at Benue State University, Makurdi, Nigeria. He obtained his BSc degree in chemistry from the same University, where he graduated *Summa Cum Laude/First Class* and Valedictorian (2009). Following a compulsory one-year National Youth Service Corps at the University of Benin, Nigeria and a year at the Benue State University as graduate assistant, Andrew proceeded to the University of Hull, United Kingdom, where he obtained his PhD degree in physical chemistry (2015) under the direction of Professor Bernard Paul Binks. His PhD thesis was *Solid Particles at Fluid Interfaces: Emulsions, Liquid Marbles, Dry Oil Powders and Oil Foams.* Andrew continued this work first as a postdoctoral fellow at the same University (2015–2016) and now as the founder of *Applied Colloid Science and Cosmeceutical Group*, Department of Chemistry, Benue State University, Makurdi, Nigeria. His research focuses mainly on particle-stabilized colloidal systems. Andrew has received over 25 accolades and has authored 50 publications: 3 books, 34 journal papers, 12 conference abstracts/proceedings and a patent. His journal papers are in top scientific journals in chemistry, including *Soft Matter, Langmuir, ACS Applied Materials and Interfaces, Colloid and Surfaces A, Journal of Dispersion Science and Technology, SN Applied Sciences* and *Journal of Colloid and Interface Science.* Andrew is on the editorial board of *Reviews of Adhesion and Adhesives* and the editor of *Frontiers in Soft Matter.* He is member of the Royal Society of Chemistry, the American Chemical Society, the Institute of Charted Chemists of Nigerian and the Chemical Society of Nigeria. Andrew lives in Makurdi, Benue State, Nigerian with his wife, Pharmacist Maryam, and three children, Aondover, Kumawuese and Aondosoo. He loves gardening and enjoys playing piano and flute.

Acknowledgments

I am thankful to the unknown reviewers for their favorable feedbacks and invaluable comments on the proposal that led to this book project. I gratefully acknowledge CRC Press staff for all they have done for the success of this book project. I am also thankful to all my colleagues and professors who took time out of their tight schedules to reply my emails and also permitted me to use one or more of their personal materials such as grant proposals, posters and sketches. I am particularly indebted to

Barbara Glunn (Former Editor, Chemistry, CRC Press/Taylor and Francis Group, LLC) who retired from publishing after commissioning me to write this book.

Alexis O'Brien (Editor, Chemistry, CRC Press/Taylor and Francis Group, LLC) who oversaw the publication of this book.

Thomas Sherry (Senior Editorial Assistant, CRC Press/Taylor and Francis Group, LLC) for sending me all the files and templates necessary for the success of this book project.

Sathya Devi (Project Manager, CRC Press/Taylor and Francis Group, LLC) for managing the production and publication of this book.

Professor Bernard Paul Binks (Department of Chemistry and Biochemistry, University of Hull, Hull, UK and Editor of Langmuir) for his mentorship during my doctoral degree and beyond.

Professor Douglas Tobias (Department of Chemistry, University of California, Irvine, US) for helping secure permission to use sample grant proposals from his university.

Professor William Tang (Samuel Department of Biomedical Engineering, University of California, Irvine, US) for permitting the use of his grant proposal.

Professor Philip G. Collins (Department of Physics and Astronomy, University of California, Irvine, US) for permitting the use of his grant proposal.

Professor Alan F. Heyduk (Department of Chemistry, University of California, Irvine, US) for permitting the use of his grant proposal.

Professor Kevin Loutherback (Rochester, MN, Laboratory Medicine and Pathology, US) for permitting the use of his grant proposal.

Professor Tracy Covey (Department of Chemistry and Biochemistry, Weber State University College of Science, US) for permitting the use of her grant proposal.

Professor Frantz Carie (Department of Earth and Environmental Sciences, Weber State University College of Science, US) for permitting the use of her grant proposal.

Professor Anderson M. William and Dr Jonathan Agger (Centre for Nanoporous Materials, University of Manchester Institute of Science and Technology, UK) for permitting the use of their grant proposal.

Professor Vernita Gordon (Department of Physics, The University of Texas at Austin, US) for permitting the use of her grant proposal.

Prof. Dr. Silvio O. Rizzoli (Center for Biostructural Imaging of Neurodegeneration, University Medical Center Göttingen, Von-Siebold-Straße 3a, 37075 Göttingen, Germany and Department of Neuro and Sensory Physiology, University Medical Center, Göttingen, Humboldtalee 23, 37073 Göttingen, Germany) for permitting the use of their short communication paper.

Professor Theoni K. Georgiou (Faculty of Engineering, Department of Materials, Imperial College London, UK) for writing the foreword.

The Vice-Chancellor Professor Joe Iorapuu and Professors Bernard Atu, Simon Ubwa, Ogbene Igbum, Mike Imande, David Tyona, Benjamin Anhwange, Slyvester Adejo, Rose Kukwa, Luter Leke, Emmanuel Mbaawuaga and Dr James Tsor (Benue State University, Makurdi, Nigeria) for their encouragement and support.

Dr Msugh Targema (Benue State University, Makurdi, Nigeria) for his friendship and painstakingly proofreading the final draft.

Vincent J. Vezza (Biomedical Engineering, University of Strathclyde, Glasgow, UK) for permitting the use of their short communication paper.

Lia A. Michaels (Department of Biological Sciences, Dartmouth College, Hanover, US) for permitting the use of her poster.

Ty J. Werdel (Horticulture and Natural Resources, Kansas State University, US) for permitting the use of his poster.

Stephanie Andersen and Laura Brossart (Brown Centre for Public Health Systems Science, Washington University in St. Louis, US) for permitting the use of their poster.

Titus Iangba (Benue State University, Makurdi, Nigeria) for his friendship, encouragement and support.

Pharmacist Maryam Terhemen (my wife), **Aondover, Kumawuese** and **Aondosoo** (children) for their love, friendship, encouragement and support.

I also thank all my students who have inspired me and I enjoy teaching.

1 Concept and Overview

Communication is the passing of information to a person (or a group of people) through either writing, speaking, singing, signs or other related means. It is a two-way process because the person or people receiving the information need(s) to understand exactly what has been passed. An effective communication takes place when the information is passed without any ambiguity. Oppositely, an ineffective communication happens when the information is ambiguous and the receiver is left to decipher its meaning. In science, where information and ideas are necessarily shared for the advancement of knowledge, effective communication is highly indispensable. Communication in science is mainly *via* writing and speaking, both of which follow a logical pattern of thoughts. Successful scientists are always strong writers and becoming a strong writer is not really a daunting task – all that is required is adherence to a set of simple rules. However, many (early career) scientists struggle with writing because universities often neglect this training, *i.e.*, modules, courses or workshops on writing are not offered. As an alternative, many early career scientists blindly mimic the writing styles of their professors or authors they came across during their training (Moore 1991). While many early career scientists have succeeded through this method, others are still struggling to gain strong writing skills to meet the ever-demanding writing need in science. This book aims to help chemical scientists, aching to gain effective writing skills achieve their dream by providing simple practical rules which many strong science writers follow. On this basis, the book takes the reader slowly through the art, providing numerous examples and follow-up activities to facilitate hands-on experience, so as to stimulate the logical thinking required in scientific writing.

1.1 SCIENTIFIC WRITING

Just like singing, dancing, carving, drawing and painting, writing (a creative form of expression) is an art and it is developed by training and practice. However, writing is a one-dimensional art that creates a multi-dimensional reality in the mind of the reader. Arts like visual (painting, photography), audio (music, spoken word) or a combination of both, audio-visual (film, theatre, performance) naturally inform more and therefore have more control over the communication process and the receiver's mind compared with writing. This notwithstanding, writing is meant to do the same thing as they do. Unlike the aforementioned arts, which are meant to entertain, writing in science is not for entertainment. A scientific write-up either disseminates information (*e.g.*, research article) or preserves a piece of information (*e.g.*, laboratory notebook) for other scientists and so precision, in terms of language usage, becomes crucial. A scientific write-up may involve a description of a process, a laboratory procedure or a description of a piece of work done on a particular area (*e.g.*, thesis, research article). Whatever the case, a sequential or step-by-step presentation of

DOI: 10.1201/9781003186748-1

ideas, *i.e.*, "logical presentation", if you like, is needed for producing an acceptable write-up. To achieve this, the writer must:

- Have a well-developed and an acceptable scientific writing style.
- Possess a good knowledge of the subject area or topic.
- Have a logical and a creative mind.
- Know the target audience.

Notice that the mastery of English language or grammar is not mentioned because good writing can be done in any language provided the necessary rules are followed. Poor English is not necessarily an excuse for poor writing, even though it adversely impacts the quality of a write-up. In fact, it is a minor issue that can be easily learned or corrected with due diligence. It is also important to note that the qualities required of a good artist are the same as those required of a good science writer. For example, a good painter has command over his/her painting. Similarly, science writers are expected to have complete command over their manuscripts even if the contents come from collaboration with others. Therefore, just like an art work is the master-piece of the artist, a scientific write-up is the masterpiece of the scientist too. That is why having a unique writing style – yet following the accepted rules – is very impor-tant. Once this is understood, *plagiarism* (copying someone else's work *verbatim*) (Park 2003) can be easily avoided.

1.2   OVERVIEW OF SCIENTIFIC WRITINGS

Scientific writings, as used here, include scientific articles or papers (*i.e.*, research articles or papers, communications, notes, reviews, perspectives, supplementary data and essays), scientific posters, research grants and book proposals, research books, theses, laboratory or fieldwork reports. Each of these has a unique structure and lan-guage style and is evaluated differently.

- *Research articles* (also called *full-length papers* or *research papers*): They give a vivid account of a completed original piece of research work. They are very important because they are source of new data and information on subject areas. They follow a rigid structure, developed hundreds of years ago. They are subjected to *peer-review* before they are accepted for pub-lication. "Peer-review" (Chapter 19) is the critical evaluation of a work by independent people with adequate experience or knowledge of the work or subject area.
- *Supplementary data*: They report experimental protocols and data gener-ated from them. They are also peer-reviewed before publication and may be published with a full-length paper or separately, *e.g.*, *Data in Brief* pub-lished by Elsevier and *F1000Research* published by Taylor and Francis.
- *Communications* (also called *letters* or *letter to editor, correspondence, rapid* or *short communications*): They are preliminary reports of special significance and urgency, and they are published as quickly as possible. They are accepted once the editor believes that their rapid publication will

benefit the scientific community, but they are also often subjected to the peer-review process. They are much shorter than full-length papers.

- *Notes*: They give precise accounts of original research work of limited coverage. They are also considered as preliminary reports of special significance, but they are not urgently needed like communications. The results reported are definitive and may not be published subsequently. Notes normally contain new techniques of interest, improved procedures of wide applicability or interest and accounts of novel observations. Notes undergo the same editorial peer-review appraisal like full-length papers. They are typically the same length as communications.
- *Reviews* (also *review articles* or *papers*): They do not report new experimental findings like full-length papers. Instead, they integrate, correlate, evaluate and critically examine results from published journal papers on a particular subject area. They may present novel theoretical interpretations of the results. They serve as pointers to the original full-length papers and as identifiers of future research direction. They are relatively very long, with plenty references, and they are subjected to peer-review before publication.
- *Perspectives*: They are similar to review articles, but they give a new or unique judgment of a fundamental concept or an existing issue. Because authors of perspective articles are expected to have in-depth knowledge of the subject area, perspective articles may also contain personal opinion(s) of the author(s) and may propose or support a new hypothesis and point future research directions.
- *Essays*: These are short pieces of write-ups on a particular topic or subject area. *Effective Writing and Publishing Scientific Papers* by Cals and Kotz (2013) and *Improve Scientific Writing and Avoid Perishing* by Carryaway (2006) are examples of published essays.
- *Research books*: They are divided into *proceedings volumes*, *monographs* and *handbooks*, with each chapter reviewed by an editor before publication. *Proceedings volumes* are divided into chapters which may contain accounts of original research or literature review. Normally, they are multi-authored with the chapters coming mainly from presentations given at symposia. Nonetheless, additional chapters that are not part of the symposium presentation are often written for the purpose of complete coverage. *Monographs* are detailed written books on a specialized single topic. They are collaboratively written by one or more authors. *Handbooks* are large volumes written in-depth on a subject area by multiple authors. Normally, the chapters are short (about three to four pages). Each chapter is written in detail by one or more authors on a narrow topic within the scope of the book.
- *Scientific posters*: A scientific poster summarizes research findings concisely and attractively. Research posters are generally not subjected to peer-review, and they are not published in scientific journals. They are used at conferences, symposia and workshops.
- *Theses*: They are an account of a personal research work on a topic written by a candidate for a higher university degree, *e.g.*, MSc or PhD. Although

the styles of writing theses differ in different countries, in many cases they follow a rigid structure akin to that of a full-length paper (Chapter 2).

- *Laboratory, fieldwork* and *industrial attachment reports*: These are reports of laboratory experimental work, fieldwork or trip and industrial placement, respectively. In many cases, these reports also follow a rigid structure similar to full-length papers.
- *Research grant proposals*: This is a written plan offering to conduct a piece of work or project. The proposal does mainly three things. First, it says who you are and why you are qualified to do the project or work. Second, it says, step-by-step, how you plan to do the project or work and the time-frame required for it. Third, it gives a detailed statement of the budget for the project or work. Research proposals always pass expert appraisal (or review) before they are awarded.
- *Book proposals*: A book proposal is written to convince a publisher to publish a book. A book proposal (i) summarizes the book's aim and selling features, (ii) gives the table of contents and marketing clues, (iii) states the book's importance and (iv) lists possible competing titles and the target audience. It may be submitted with sample chapters (if available). Book proposals must pass expert review before a formal contract is drawn between the publisher and the author(s).

Given the broad genre of scientific writings, only peer-reviewed scientific articles (research papers, short communications, review articles and perspectives), scientific posters, oral presentations, research grants and book proposals will be covered in this book. Experience gained from these can be used to write other scientific genres.

1.3 WHY IS WRITING AND PUBLISHING IMPORTANT TO SCIENTISTS?

Writing and publishing are important to scientists:

- For the advancement of knowledge in their field.
- For obtaining higher degrees, *e.g.*, some institutions require publications for the award of doctoral degrees.
- For career growth, *e.g.*, it is impossible to become a university professor without publications.
- For fame and recognition (*i.e.*, many scientists want to be famous and to be recognized for their contribution in a field).

Irrespective of these reasons, publishing in high-impact factor scholarly peer-reviewed journals is the "Holy grail" of all scientists. However, some resort to predatory journals (*i.e.*, journals that aim at making money from authors, and therefore do not sufficiently scrutinize their submitted work before publishing (Beall 2012)) after repeated rejection from the high-impact factor journals, especially when they are due for promotion so as to meet the required number of journal articles.

1.4 CHARACTERISTICS OF SCIENTIFIC WRITINGS

Scientific writings are *precise*, *impersonal* and *objective*. Typically, they contain the third-person language, passive and active tenses, with complex terminologies, footnotes and references.

- *Precision*: Scientific writings convey information without any ambiguity. Ideas are expressed in the simplest possible ways. Irrelevant expressions, interjections, rambling sentences and excessive words are absent. Nonetheless, no important details are left out. For example, Figure 1.1 contains two images of a football. In terms of scientific writing, it is wrong to think of these images as "real footballs". Figure 1.1a is a photograph of a football, while Figure 1.1b is a cartoon of the same football. Read the figure caption and note how the images are described precisely. This is typically how precision works in scientific writing.
- *Objectivity*: Scientific writings present a balanced range of views on the subject area under consideration based on experimental data, leaving out views relating to moral briefs of what is "right" or "wrong".
- *Impersonality*: Scientific writings are written mainly in third person, and as such, the use of personal pronouns is limited. This is because, without say, it is understood to be the author's work. The use of personal pronouns is allowed when an opinion is attributed to someone else. Therefore, the use of phrases like "in my opinion", "I think", "I am convinced that" and "I understand" are unnecessary. Plural pronouns like "we" and "our" are also not used for the same reasons. Notwithstanding, some journals accept the use of personal pronouns like "I", "We" and "Our" as shown in the textual abstract of Hijnen and Clegg (2014) (Example 1.1) where the first-person plural pronoun "We" is used.

Example 1.1 *Use of Pronouns in a Scientific Paper*

"Colloidal particles are often regarded as building blocks for creating new materials, so guiding them to form specific structures is crucial. Here we present a simple method to assemble colloids into a cellular network. It relies on compositional changes in a colloid-stabilized emulsion of partially miscible liquids, and is induced by the disappearance of the continuous phase *via* evaporation. The particles are hereby forced into a cellular network as the droplets are squeezed together. Re-mixing of the liquid phases eventually occurs, leaving a stable structure surrounded by a single fluid phase. We demonstrate in detail the formation of the networks and discuss some general features of this approach. Finally, we show the macroscopic volume change and ultimate stability of the structure".

From Hijnen and Clegg (2014, *Mater. Horiz.* **1** (3), 360-364)

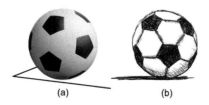

(a)   (b)

FIGURE 1.1   (a) A photograph and (b) a cartoon of a football, with white and black patches, in a white background.

Also, the world-changing paper of Watson and Crick (1953) on the structure of DNA began with the first-person plural pronoun – "We wish to suggest a structure for the salt of deoxyribose nucleic acid (DNA)". According to Kuo (1999), the use of first-person plural pronoun is more common than other pronouns. When used, the authors often refer to themselves alone or themselves and their co-workers (Kuo 1999).

Activity 1.1 *Identifying Pronouns in Excerpt of Scientific Papers*

Circle the personal pronoun in the following excerpt:

a. "However, aggregates, complexes and coacervates are obviously particles that are complex in terms of their structure and composition. In this work we describe three types of particles of fairly simple structure and composition that are definitely compatible with foods and how these interact with proteins in foams and an oil-in-water emulsion. The particles are: (i) stable, non-spreading oil droplets, (ii) hydrophobically-modified cellulose fragments and (iii) hydrophobically-modified starch granule particulates".

From Murray et al. (2011, *Food Hydrocoll.* **25** (4), 627–638)

b. "Our language idealizes protein-protein interactions, essentially as a particular restricted kind of graph-rewriting operating on graph-with-sites not unlike Lafont's interaction nets. Bindings are explicit: a formal protein is a node with a fixed number of sites, a complex is a connected graph built over such nodes".

From Danos and Laneve (2004, *Theor. Comput. Sci.* **325** (1), 69–110)

Activity 1.2 *Identifying Pronouns in Scientific Papers*

Examine at least five papers from reputable journals in your field for the use of personal pronouns.

1.5 WRITING SCIENTIFICALLY

Writing generally has three main elements, namely *words*, *sentences* and *paragraphs*. Poor selection of words, poor construction of sentences and paragraphs mar the goal of any write-up, especially in science. This section gives a few general suggestions on how to choose the right words and phrases and combine them into sentences and paragraphs to form a good scientific write-up. More resources on word usage and grammar are available in Appendix I.

- *Words*: In scientific writing, only the ability to communicate clearly is needed. Therefore, avoid superfluous words and word wastage. Use words with precision and economy to construct phrases and sentences that convey an idea in a simple and clear manner. Table 1.1 contains examples of superfluous use of words in comparison to their precise usage.

  Excessive use of *jargon* can also impede understanding, and their usage must be kept to a minimum. The general rule is to avoid the use of new words or phrases, especially if there are already existing ones for the intended purpose. For the same reason, excessive use of abbreviations and acronyms should be avoided. If you must use abbreviations, make them meaningful. For example, rather than use k_1 and k_2 to symbolize the *forward* and *backward* rate constants, respectively, of a reaction (Equation 1.1), use k_f and k_b, respectively, as readers are more likely to remember them without looking up their meaning repeatedly compared with k_1 and k_2.

$$A + B \underset{k_b}{\overset{k_f}{\rightleftharpoons}} P$$

(1.1)

  Similarly, when using identifiers, use names logical enough for the readers to naturally keep track of. For example, "protein dilution buffer" and "4-amino-1,10-phenanthroline" are far easier to understand and keep track of than say "buffer A" or "compound A or 2". Alternatively, make a table with a list of abbreviations, acronyms as well as symbols. This is especially important in relatively long review papers.

TABLE 1.1
Examples of Superfluous and Precise Use of Words in Scientific Writing

Superfluous	Precise
… in order to	… to
… by means of	… by, with, via
… if conditions are such that	… if
… there can be little doubt that this is	… this probably is
… created the possibility	… made possible
… due to the fact that	… because
… fewer in number	… few
… for the reason that	… since, because
… in all cases	… always
… it would appear that	… apparently
… because of the fact that	… because
… bright red in color	… bright red
… not possible	… impossible

- *Tenses* also impact the quality of a scientific write-up. Scientific observations are best reported in past tense, but majority of them are reported in present tense because the present tense is more engaging than the past tense. The present tense, for instance, is used when the experiments can be reproduced with consistent observations as shown in Example 1.2 by Tyowua et al. (2017). In past tense, the sentence could have been written as "The two fumed silica particle types preferred to stabilize different emulsion types upon shearing equal volumes of vegetable oil and 100 cS silicone oil". This sentence is less engaging compared with the one in Example 1.2.

Example 1.2 *The Use of Present Tense in Scientific Writing*

"The two fumed silica particle types prefer to stabilize different emulsion types upon shearing equal volumes of vegetable oil and 100 cS silicone oil".
        From Tyowua et al. (2017, *J. Colloid Interface Sci.* **488**,127–134)

Activity 1.3 *The Use of Past and Present Tenses in Scientific Papers*

Analyze at least five papers in your field from reputable scientists for the use of past and present tenses.

As a guide:

a. Use the present tense for known facts and hypotheses. For example, "pure water boils at 100°C" (*known fact*). "The purpose of this study is to test HPLC-separated Aloe vera gel components against a comprehensive panel of microbes to characterize their antimicrobial activities" (Cock 2008) (*hypothesis*).
b. Use the past tense to refer to conducted experiments. For example, "a simple Wilhelmy cycle containing an advancing phase and a receding phase was used to measure the surface tension of water" (Arbatan and Shen 2011).
c. Use the past tense to describe results peculiar to your experiments. For example, "the water sample boiled at 90°C". This result is peculiar to the experiment because pure water boils at 100°C.

- *Sentences*: According to the *Oxford English Dictionary*, a sentence is a series of words in connected speech or writing that forms a grammatically complete expression of a single thought. Types of sentences and their grammatical structure can be learned in standard English language grammar textbooks or some of the materials suggested here for further reading. Structurally, complete sentences always contain a *subject*, a *verb* and an *object*. The subject performs the action; the verb describes the action of the subject while the object receives the action of the verb. The verb is the most important part of a sentence because it can even exist with the subject alone to convey meaning, *e.g.*, "the liquid is boiling" is a meaningful sentence containing the subject "the liquid", the verb be "is" and the infinitive verb "boiling". Another importance of a verb in a sentence is that it must agree with the subject for the sentence to be grammatically correct. The verb plus the object forms the *predicate* (*i.e.*, a group of words that describe the action of the subject), but the predicate is meaningless without the subject. For example, with the object "the beaker", the sentence becomes "the liquid is boiling in the beaker", where "is boiling in the beaker" is the predicate, which is meaningless without the subject. A sentence is either in an *active voice* or a *passive voice*. In an active voice, the subject, the verb and the object follow themselves sequentially, *i.e.*, the subject *performs* the action described by the verb and the object receives the action of the verb. On the contrary, the action comes before the subject in a passive voice, *i.e.*, the subject *receives* the action of the verb. For example, "James et al. (2022) prepared the aqueous salt solution" is an active sentence with the subject "James et al. (2022)", which is followed by the verb "prepared" and then the object "the aqueous salt solution". In the passive voice, the sentence will read "the aqueous salt solution was prepared by James et al. (2022)". In this case, the object comes first, followed by the verb (*i.e.*, the action) and then lastly by the subject. Scientific writings usually contain a mixture of sentences in both the active and passive voices, but the active voice is more beneficial and, therefore, it should be largely used. Active sentences are more precise, more understandable, clearer and more persuasive than their passive counterparts. Scientists use passive sentences for three reasons: (i) when the doer of the action is unknown, *e.g.*, "the surface was coated with copper" (passive, doer unknown) *cf.* "Jones et al. (2000) coated the surface with copper" (active, doer known); (ii) when it is not important to name the doer of the action in the sentence, *e.g.*, "the surface was coated with copper (Jones et al. 2000)" (doer known, not part of the sentence); and (iii) when they do not want to personalize their action, *e.g.*, "the surface was coated with copper" (action not personalized) *cf.* "we coated the surface with copper" (action personalized). The problem with passive sentences is that it is sometimes difficult for readers to know when the authors are not personalizing their action and when they are ascribing the action to others (*e.g.*, *cf.* the passive sentences in (i) and (iii)). In such an instance, the active sentences are more preferable. Moreso, with many scientific journals and readers now accepting the active voice, the use of passive voice in scientific writings is fading.

As a guide:

a. Use the active voice to report and describe research findings and attribute arguments and conclusions to their authors. For example, "Tanaka et al. (1994) reported pattern evolution in phase-separating polymer blends in the presence of glass beads" (Thijssen and Clegg 2010), and "Wu et al. (2008) fabricated directionally aligned poly(vinyl butyral) nanofiber arrays by electrospinning" (Shin et al. 2016). The active voice is also used for presenting different sides of an argument and for conveying messages that are direct and to the point. For example, the sentence "pure palm oil is red while pure water is colorless" presents two sides of an argument: one saying pure palm oil is red, and the other saying pure water is colorless.

b. Use the passive voice to (i) report experimental procedures, (ii) state general knowledge, (iii) give information that is widely accepted and/or known to be true and (iv) give pieces of information or evidence that supports the same conclusion. Here are few examples, "superhydrophobic surfaces were manufactured in a way similar to that previously reported by the authors" (Bormashenko et al. 2006) (*experimental procedure*), "the natural habitant of a fish is water" (Jones 2018) (*general knowledge*), "obesity is a serious health problem" (Ogben 2007) (*widely accepted and known to be true*) and "evidence from optical microscope and scanning electron microscope show that the emulsions are made up of spherical oil droplets" (Tyowua 2018) (*pieces of evidence that support the same conclusion*). Note that these sentences are passive because they do not contain the authors' names. However, the authors are mentioned in parenthesis for reference's sake.

c. Vary sentence length, but avoid excessive use of short and long sentences. The contents of short sentences (≤ 20 words) are easier to understand and remember than the contents of long ones (≥ 20 words) (Miller and Selfridge 1950). However, average sentence length in published scientific literature is 25–30 words (Moore 2011); so, keep your sentences a few words below or above this range.

d. Ensure that the subject, verb and object are close together.

Activity 1.4 *The Use of Active and Passive Sentences in Scientific Papers*

Identify the active and passive sentences in the following excerpt:

"Emulsions and foams can be stabilized not only by surfactants, but also by solid particles. Emulsions of this type are termed Pickering emulsions, after the early work of Pickering (1907) although essentially the same phenomena were sketched out earlier by Ramsden (1903). Binks and Horozov (2006) have provided a survey of the subject of particle-stabilized colloids in general. The stabilization of food foams by particles has been reviewed by Murray and Ettelaie (2004) whilst Dickinson (2010) has more recently reviewed the stabilization of both emulsions and foams by particles in the context of foods. The key feature of particle-stabilized systems is that if the particles have the correct surface energy or contact angle with the interface, and also that they have a sufficiently large surface area, then the energy of desorption per particle can be of the order of several thousand kT. Such particles are thus effectively irreversibly adsorbed".

From Murray et al. (2011, *Food Hydrocoll.* 25 (4):627–638)

Activity 1.5 *Sentence Length in Published Journal Papers*

Count the number of words per sentence in at least five papers in your area. Prepare a frequency table and draw a bar chart or histogram of the frequency of number of words per sentence versus the number of words per sentence. On average, what are the minimum and maximum numbers of words per sentence? What is the number of words in majority and minority of the sentences?

- *Paragraphs.* A paragraph is a unit of thought. Each paragraph contains a topic sentence that forms the theme of the paragraph. All other sentences support this theme. There is no general rule to the length of a paragraph, but a paragraph is supposed to contain at least five sentences. A good paragraph is logical and the sentences overlap, *i.e.*, the first sentence gives an idea of what the second sentence would be and so on. In addition, a good paragraph is understandable even when it stands alone.

As a guide:
a. Use the first sentence of your paragraph as the topic sentence to tell the reader what to expect from the paragraph. Thereafter, use the rest of the sentences to support the first.
b. The first sentence of your paragraph should not refer to an idea in the preceding paragraph. To do this, avoid starting a paragraph with words and phrases like "However", "Nonetheless", "Its", "These", "They" and "In contrast" and similar words or phrases. If you must start like this, use constructs that will naturally tell the reader that the paragraph is related to the preceding one. For example, you can say "In support of the above arguments…", "In addition to these …" (Plaxco 2010).
c. Each sentence should arise logically from the one preceding it. By doing this, you lead the reader by the hand rather than leaving them to struggle and find out how an idea in a sentence is linked to the one preceding it. This can be achieved by using appropriate transition or connecting elements. There are three types of connecting elements, namely those that can be classified under "And" (*i.e.*, addition and result), "Or" and "But".
   i. The "*and*" connectors: The "and" connecting elements include "Moreover", "Moreso", "Furthermore", "In addition", "More importantly" and "Above all". *E.g.*, "a typical ultra-filtration system is normally made up of an ultra-filtration membrane, pump(s), reservoir, tubing, fittings, a container for collecting the permeate solution and *in addition* a pre-filtration system". "Furthermore", "More importantly" and "Above all" will also work in the sentence. The "result" connectors include "Therefore", "Because of this…" "As a result", "Thus", "Accordingly" and "Consequently". *E.g.*, "the particles are hydrophilic and *therefore* it is necessary to coat them with dichlorodimethylsilane to make them hydrophobic". Another way is to divide the sentence into two to read "The particles are hydrophilic. *Therefore*, it is necessary to coat them with dichlorodimethylsilane to make them hydrophobic".

ii. The *"or"* connectors: These include "In other words", "To put it more simply" and "That is (also *i.e.*)", "That is to say", "Or rather", "Or, more precisely". *E.g.*, "the particles are hydrophobic, *that is* they do not interact with water". The sentence can also be divided into two to read "The particles are hydrophobic. *In other words*, they do not interact with water".

iii. The *"but"* connectors: Used for conceding a point and making counter arguments. These include "However", "Nonetheless" and "Nevertheless". *E.g.*, "the small solid particles were dispersible in oil. *However*, they did not disperse in water".

Activity 1.6 *Identification of Connectors*

List all the connectors in the excerpt of Activity 1.4.

A paragraph is typically structured as illustrated in Figure 1.2. First, there is a sentence carrying the central idea. This is followed by a sentence or other sentences that support(s), restate(s) or make(s) counterpoints about the central idea. Finally, example(s), reason(s) and/or consequences are given.

Examine how the paragraph in Example 1.3, taken from Tyowua (2017), is structured and see that the same standard is held. Notice that the paragraph begins with a topic sentence having "surface tension measurement" as the central idea. The next sentence is an additional statement about the central idea (surface tension measurement). Lastly, an example is given, with reason, to support the additional statement. Also, examine all the paragraphs in this book and see how they follow the same structure.

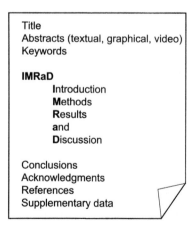

FIGURE 1.2   An illustration of the structure of a paragraph. There is a statement of the central idea in the topic sentence. This is followed by an additional statement, alternative statement or a limitation of the idea. Thereafter, an example, reason or result is given. All of these are meant to support the central idea embodied in the topic sentence.

Example 1.3 *Structuring a Paragraph*

"Surface tension is a fundamental liquid parameter in the investigation of fluid interfaces and thus its measurement is highly indispensable. Many methods (classical and modern) for measuring its value are available and the choice of the method depends on a given system. For instance, the surface tension of the interface between a static, aqueous solution and air can be measured by a number of methods, but if the solution is viscous or the surface tension is changing rapidly; care must be taken to choose the right technique".
From Tyowua (2017, Modern Principles of Colloid and Surface Chemistry, p. 101)

As a guide, when writing:

- Go straight to the point, bearing in mind that you are telling a story of your research; active sentences are especially important here.
- Report things clearly and as concisely as possible. By doing so, the use of unnecessary words or phrases can be avoided.
- Make the reader's job in terms of understanding as easy as possible: do not assume that the reader knows the research work like you.

FURTHER READING

Alley, M. 1996. *The Craft of Scientific Writing*: Springer, New York.
- Discusses the essential ingredients of scientific writing with emphases on the target audience, purpose of the write-up, precision of expressions, content organization, proper use of words and tenses as well as other aspects of grammar.

Bailey, S. 2014. *Academic Writing: A Handbook for International Students*: Routledge, New York.
- Offers an excellent guide on grammar, sentence structure and punctuation as well as on critical reading and writing.

Coghill, A.M., and L.R. Garson. 2006. *The ACS Style Guide: Effective Communication of Scientific Information*: Oxford University Press, New York.
- Discusses scientific writing in chemistry in terms of genre, grammar, sentence structure, punctuation and referencing.

Lester, J.D., and J.D. Lester. 2012. *Writing Research Papers: A Complete Guide*: Pearson, Essex.
- An excellent compendium containing a great introduction to academic writing and useful suggestions for surfing the Internet, organizing ideas, using library resources, conducting research, reading published materials, avoiding plagiarism, referencing styles as well as on writing, revising, formatting and proofreading drafts.

Strunk Jr, W., and E.B. White. 2007. *The Elements of Style Illustrated*: Penguin, New York.
- Gives an excellent guide on grammar and sentence structure with respect to academic writing.

Wallwork, A., and A. Southern.2020. *100 Tips to Avoid Mistakes in Academic Writing and Presenting*: Springer, Cham.

- It is loaded with general tips on writing as well as tips on grammar, punctuation and sentence structure.

Wiley-Blackwell House Style Guide. *House_style_guide_ROW4520101451415.pdf (wiley.com).*

- Contains excellent suggestions on voice and tense usage as well as other aspects of grammar, writing, formatting and editing.

REFERENCES

Arbatan, T., and W. Shen. 2011. Measurement of the surface tension of liquid marbles. *Langmuir* 27 (21):12923–12929.

Beall, J. 2012. Predatory publishers are corrupting open access. *Nature* 489 (7415):179–179.

Binks, B.P., and T.S. Horozov. 2006. *Colloidal Particles at Liquid Interfaces*: Cambridge University Press, Cambridge.

Cals, J.W.L., and D. Kotz. 2013. Effective writing and publishing scientific papers, part II: Title and abstract. *J. Clin. Epidemiol.* 66 (6):585.

Carryaway, L.N. 2006. Improve scientific writing and avoid perishing. *Am. Midl. Nat.* 155 (2):383–394, 12.

Danos, V., and C. Laneve. 2004. Formal molecular biology. *Theor. Comput. Sci.* 325 (1):69–110.

Dickinson, E. 2010. Food emulsions and foams: Stabilization by particles. *Curr. Opin. Colloid Interface Sci.* 15 (1–2):40–49.

Hijnen, N., and P.S. Clegg. 2014. Assembling cellular networks of colloids via emulsions of partially miscible liquids: A compositional approach. *Mater. Horiz.* 1 (3):360–364.

Kuo, C.-H. 1999. The use of personal pronouns: Role relationships in scientific journal articles. *English for Specific Purposes* 18 (2):121–138.

Miller, G.A., and J.A. Selfridge. 1950. Verbal context and the recall of meaningful material. *Am. J. Psych.* 63 (2):176–185.

Moore, A. 2011. The long sentence: A disservice to science in the Internet age. *BioEssays* 33 (12):193–193.

Moore, R. 1991. How we write about biology. *Am. Biol. Teach.* 53 (7):388–389.

Murray, B.S., and R. Ettelaie. 2004. Foam stability: proteins and nanoparticles. *Curr. Opin. Colloid Interface Sci.* 9 (5):314–320.

Murray, B.S., K. Durga, A. Yusoff, and S.D. Stoyanov. 2011. Stabilization of foams and emulsions by mixtures of surface active food-grade particles and proteins. *Food Hydrocoll.* 25 (4):627–638.

Park, C. 2003. In other (people's) words: Plagiarism by university students--literature and lessons. *Assess. Eval. High. Educ* 28 (5):471–488.

Pickering, S.U. 1907. CXCVI.-Emulsions. *J. Chem. Soc.* 91:2001–2021.

Plaxco, K.W. 2010. The art of writing science. *Protein Sci.* 19 (12):2261.

Ramsden, W. 1903. Separation of solids in the surface-layers of solutions and 'suspension'-preliminary account. *Proc. Roy. Soc.* 72:156–164.

Shin, S., J. Seo, H. Han, S. Kang, H. Kim, and T. Lee. 2016. Bio-inspired extreme wetting surfaces for biomedical applications. *Materials* 9 (116):1–26.

Thijssen, J.H.J., and P.S. Clegg. 2010. Demixing, remixing and cellular networks in binary liquids containing colloidal particles. *Soft Matter* 6 (6):1182–1190.

Tyowua, A.T. 2017. *Modern Principles of Colloid and Surface Chemistry*. Lagos, Nigeria: Academy Press.

Tyowua, A.T., S.G. Yiase, and B.P. Binks. 2017. Double oil-in-oil-in-oil emulsions stabilised solely by particles. *J. Colloid Interface Sci.* 488:127–134.

Watson, J.D., and F. Crick. 1953. A structure for deoxyribose nucleic acid. *Nature* 171 (3):737–738.

2 Getting Ready

2.1 THE RESEARCH STAGE

Research in science is all about searching for the truth. Therefore, experiments are always aimed at verifying existing knowledge, debunking existing knowledge or searching for new knowledge. This ultimately leads to new inventions or solutions to particular problems for the benefit of humanity. Irrespective of what you have set out to achieve:

- Begin your research with a thorough literature search so that you are well-informed about current issues in the area and also equipped with the relevant literatures relating to your work. Additionally, this will help you in writing the introductory part of the paper or deciding the sections if planning a review paper.
- Design your experiments with the aim to publish – this normally comes with experience, but the literature you have garnered will serve as a guide.
- Carryout your experiments, recording every protocol and outcome – these become the skeletal framework of the paper.
- Process your results and findings into graphs, tables, images (or photographs) – this normally helps with creating the storyline.

2.2 MANUSCRIPT PLANNING STAGE

Once the storyline has been created, check the aims and scope of relevant journals in the area and decide which one to write the paper for. You may have a very good story that will interest the scientific community of the area and yet be rejected if the aims and scope of the journal are different than those of your work. Find the *Guide for Authors* of the journal whose aims and scope overlap with the work you want to publish, and which you have also decided to the write the paper for, and read it thoroughly. Pay special attention to the prescribed structure and reference style. Although all scientific research papers have the same structural elements (Figure 2.1), the structural format differs across journals and you risk rejection once your paper deviates from the prescribed format.

For example, the journal *Nature* prescribes that the method section comes after the introduction, results and discussion sections, while *Langmuir* and *Journal of Colloids and Interface Science* prescribe the format given in Figure 2.1. Generally, the **IMRaD** structural format (Wu 2011) is the commonest. Nonetheless, the structure is not immediately obvious in some short communications and notes. The IMRaD structure is also abbreviated as **"IMReD"** to mean Introduction, Methods, **Re**sults and **D**iscussion or extended to include the abstracts **"AIMRaD"** to mean

DOI: 10.1201/9781003186748-2

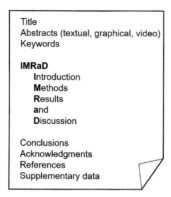

FIGURE 2.1   Structure of a scientific research paper, showing the typical elements and the IMRaD structural format.

Abstracts, **I**ntroduction, **M**ethods, **R**esults **a**nd **D**iscussion. Do not be distracted by the ever-changing variations of the IMRaD structure. The critical thing is to include these sections clearly in your paper *vis-á-vis* the guide for authors.

Activity 2.1 *Identifying Various Sections of a Scientific Research Paper*

Examine at least five research papers from different reputable journals in your field for the IMRaD structure and its variants.

2.3   MANUSCRIPT PREPARATION STAGE

One obvious question is which section to begin writing: the title, abstract (textual, graphical or video), keywords, introduction, methods, results and discussion, conclusions or the references? People start differently. Hartley (1999) suggests starting with the textual abstract and developing it into the full paper. Moreira and Haahtela (2011) recommend going from the method, results, introduction, discussion to the textual abstract and then the title. I normally begin with the **"Results"**, processing them into graphs, tables and images and then I move to the **"Experimental"** section (also called *Materials and Method*) to describe how the results were obtained. Next, I sketch out the **"Discussion"** section, explaining the results and their ramifications. I then sequentially write the **"Introduction"**, the **"Conclusions"**, the **"Abstracts (textual, graphical** and **video)"**, the **"Keywords"** (Figure 2.2), the names of the authors and their affiliations. I then decide the "Title" of the paper. Finally, I write the "Acknowledgments". The **"References"** are generated automatically if I am using a referencing software like Endnote, Zotero, Mendeley or RefWorks, but I edit them before submitting the paper. The guiding principles are (i) start with the easiest section and (ii) write in a circular manner (Figure 2.2), *i.e.*, "write and re-write" the sections. Once finished, read the paper from the title to the end, ensuring that all the sections flow together before submitting to the journal. A detailed guide for preparing the title, abstracts, keywords, introductory, experimental, results, discussion and conclusions sections of a scientific research paper is given in the next chapters.

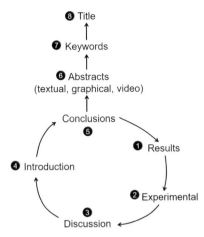

FIGURE 2.2   Schematic illustration of the path for writing a research paper with the results section as the starting point.

FURTHER READING

Cargill, M., and P. O'Connor. 2013. *Writing Scientific Research Articles: Strategy and Steps*: John Wiley & Sons, West Sussex.
- Discusses the IMRaD structure of scientific research papers, including its variations, and gives useful suggestions for writing the various sections of the paper.

Dominiczak, M.H. 2013. Art, science, words, and IMRAD. *Clin. Chem.* 59 (12):1829-1831.
- Discusses the history of the IMRaD structure of scientific research papers.

REFERENCES

Hartley, J. 1999. From structured abstracts to structured articles: a modest proposal. *J. Tech. Writ. Commun.* 29 (3):255–270.

Moreira, A., and T. Haahtela. 2011. How to write a scientific paper – and win the game scientists play! *Pulmonology* 17 (3):146–149.

Wu, J. 2011. Improving the writing of research papers: IMRAD and beyond. *Landscape Ecol.* 26 (10):1345–1349.

3 Research Paper Title

3.1 FUNCTIONS OF THE TITLE

The primary function of the title is to tell readers what the paper is all about and further entice them to read it (Haggan 2004). Additionally,

- prior to publication, editors read the title along with the textual abstract to determine whether the manuscript is suitable for their journal or not and
- after publication, the title is indexed along with the textual abstract and authors' information to enable retrieval of the paper in online searches (Cals and Kotz 2013).

3.2 TYPES OF TITLES

Titles are either *descriptive/neutral*, *declarative/informative* or *interrogative*.

- Descriptive titles state the subject and design of the study but do not disclose the findings of the study (Cals and Kotz 2013), thereby raising the reader's curiosity. *Advantages*: (i) Descriptive titles cause the reader to delve more into the paper for insight, and (ii) they allow the reader to interpret the findings open-minded without bias. *Disadvantages*: Descriptive titles have little information and may mislead the reader to wrong conclusions.
- Declarative titles disclose the findings of the study (Cals and Kotz 2013), thereby reducing the curiosity of the reader. *Advantage*: The reader does not need to read the entire paper to get the central message. *Disadvantage*: Declarative titles raise bias and do not allow the reader to interpret the findings objectively.
- Interrogative titles have a query or a question formulated around the research question but do not reveal the findings. *Advantages*: (i) They emphasize the research question or the theme of the paper, and (ii) they raise curiosity in readers' minds and cause them to read the paper for answers to the question posed in the title. *Disadvantage*: May cause the reader to miss other important information contained in the paper while searching the paper for answer(s) to the question posed in title.

3.3 GRAMMATICAL STRUCTURE OF TITLES

Titles are grammatically structured into *nominal*, *full-sentence*, *compound* and *question* (Soler 2007).

DOI: 10.1201/9781003186748-3

- Nominal titles do not have the structure of a sentence – *i.e.*, the subject, verb and predicate. Nominal titles are common in both research and review papers (Soler 2007). Here are some examples: "Water splitting-biosynthetic system with CO_2 reduction efficiencies exceeding photosynthesis" (research paper) (Liu et al. 2016) and "Artificial photosynthesis for hydrogen production" (review paper) (Chen et al. 2019).
- Full-sentence titles have the structure of a complete sentence and contains a subject, a verb and a predicate. Full-sentence structures are useful for titles that disclose the findings of the paper. Full-sentence titles are often used in research papers but rarely used in review papers (Soler 2007). Some examples include: "Artificial leaf turns sunlight into a cheap energy source" (research paper) (Service 2011) and "Carbon nanotubes are able to penetrate plant seed coat and dramatically affect seed germination and plant growth" (research paper) (Khodakovskaya et al. 2009).
- Compound titles are divided into two interrelated parts by a colon or a dash, *e.g.*, two noun phrases separated by a colon (Haggan 2004). Compound titles are common in both research and review papers (Soler 2007). Examples are "Honey: single food stuff comprises many drugs" (review paper) (Khan et al. 2018) and "Cannabis, the pregnant woman and her child: weeding out the myths" (review paper) (Jaques et al. 2014).
- Question titles are interrogative in nature. They are rarely used in both research and review papers (Soler 2007). Examples include "Cassie and Wenzel: were they really so wrong?" (research paper) (McHale 2007) and "Why does the Cassie-Baxter equation apply?" (comment) (Bormashenko 2008).

3.4 WRITING THE TITLE

Some journals give specifications of acceptable titles. Majority of journals allow single titles, whereas others require compound titles which contain two titles – a main title and a subtitle separated from each by a colon. Some journals require a short running title to place at the top or at the bottom of pages to facilitate readers' navigation of the journal (Cals and Kotz 2013). Whatever the case may be, your title should be (Hartley 2008):

- *Attractive* – lure the reader to the paper.
- *Informative* – make clear the content of the paper.
- *Accurate* – concise in context of subject area.
- *Short* – not more than 20 words if possible; papers with short titles are more read and cited than those with long titles (Letchford et al. 2015).
- *Devoid of abbreviations* and *acronyms* except those that are well-known like "pH".

The following strategies can be used to create attractive, informative, accurate and short titles:

- Be concise, but include important information as much as possible.
- Place keywords that will arrest readers' attention at prominent positions, *e.g.*, at the beginning or near the beginning of the title; one way to create a keywords-prominent title is to use the compound structure.
- Decide between noun phrase, statement or question structure, but be sure to avoid ambiguous phrases and constructions.
- Be sure not to begin your title with the following vague phrases: "*Effect(s) of …*", "*Investigation of …*", "*Determination of …*", "*On the …*" and other similar ones.

Activity 3.1 *Classifying Paper Titles into Groups*

Examine the title of ten papers in your area and classify them into the various groups as discussed in this chapter.

FURTHER READING

Parija, S.C., and V. Kate, eds. 2017. *Writing and Publishing a Scientific Research Paper*: Springer, Singapore.
- Discusses general aspects of writing and publishing a scientific research paper, including useful suggestions for creating a meaningful title for the paper.

Richardson, S.M., F. Bella, V. Mougel, and J.V. Milić. 2021. Scientific writing and publishing for early-career researchers from the perspective of young chemists. *J. Mater. Chem. A* 9 (35):18674–18680.
- Contains useful suggestions for writing various sections of a scientific research paper, including the title.

REFERENCES

Bormashenko, E. 2008. Why does the Cassie–Baxter equation apply? *Colloids Surf. A* 324 (1):47–50.

Cals, J.W., and D. Kotz. 2013. Effective writing and publishing scientific papers, part II: Title and abstract. *J. Clin. Epidemiol.* 66 (6):585.

Chen, Y., X. Li, C. Tung, and L. Wu. 2019. Artificial photosynthesis for hydrogen production. *Prog. Chem.* 31:38–49.

Haggan, M. 2004. Research paper titles in literature, linguistics and science: Dimensions of attraction. *J. Pragmat.* 36 (2):293–317.

Hartley, J. 2008. *Academic Writing and Publishing: A Practical Handbook*: Routledge, New York.

Jaques, S.C., A. Kingsbury, P. Henshcke, C. Chomchai, S. Clews, J. Falconer, M.E. Abdel-Latif, J.M. Feller, and J.L. Oei. 2014. Cannabis, the pregnant woman and her child: weeding out the myths. *J. Perinatol.* 34 (6):417–424.

Khan, S.U., S.I. Anjum, K. Rahman, M.J. Ansari, W.U. Khan, S. Kamal, B. Khattak, A. Muhammad, and H.U. Khan. 2018. Honey: single food stuff comprises many drugs. *Saudi J. Biol. Sci.* 25 (2):320–325.

Khodakovskaya, M., E. Dervishi, M. Mahmood, Y. Xu, Z. Li, F. Watanabe, and A.S. Biris. 2009. Carbon nanotubes are able to penetrate plant seed coat and dramatically affect seed germination and plant growth. *ACS Nano* 3 (10):3221–3227.

Letchford, A., H.S. Moat, and T. Preis. 2015. The advantage of short paper titles. *R. Soc. Open Sci.* 2 (8):150266.

Liu, C., B.C. Colón, M. Ziesack, P.A. Silver, and D.G. Nocera. 2016. Water splitting–biosynthetic system with CO_2 reduction efficiencies exceeding photosynthesis. *Science* 352 (6290):1210.

McHale, G. 2007. Cassie and Wenzel: were they really so wrong? *Langmuir* 23 (15):8200–8205.

Service, R.F. 2011. Artificial leaf turns sunlight into a cheap energy source. *Science* 332 (6025):25.

Soler, V. 2007. Writing titles in science: An exploratory study. *English for Specific Purposes* 26 (1):90–102.

4 Authorship, Acknowledgments, Competing Interests and References

4.1 AUTHORSHIP

For many journals, the name(s) of author(s) as well as their addresses and affiliation(s) come immediately after the title while the acknowledgments, competing interests and reference sections come at the end of the paper. Nonetheless, these are treated together in this chapter because they share some overlapping characteristics, *e.g.*, all of them have an overall contributory integrity on the published paper. Deciding between who should be and who should not be an author as well as the order of author names, for multi-authored papers, are delicate issues among researchers (Brand et al. 2015). These issues often lead to dispute and also discourage future collaboration. Based on the *Guide for Authors* of many scientific journals, an author is someone who has made *significant contribution* to the work being published. However, what can be considered a significant or an insignificant contribution? The answer to this question is subjective: a contribution may be considered significant by one and considered insignificant by another. Therefore, this leaves the lead researcher or team leader with the onerous tasks of deciding the authors and their order of names. The real issue is that a fair decision is not always made and others are falsely denied authorship status. The silver lining is that no journal stipulates the maximum number of authors per paper. In fact, papers with more than a thousand authors have been published (Brand et al. 2015, Teodosiu 2019). Therefore, authorship status can be given to all who have contributed in one way or the other to the work being published. This will undoubtedly resolve the dicey issue of authorship, but the issue of name order remains. That is, of these authors, whose name comes first, second, third, etc.?

The order of names for multi-authored papers is often decided by the lead researcher or team leader, but this is always a contentious issue (Brand et al. 2015). This is contentious because of the wrong perception that the order of names reflects the degree of contribution of the authors to the work. To resolve this, some journals and research group use alphabetical order of surname to list authors' names (Brand et al. 2015). This means those whose surnames begin with earlier alphabets occupy the initial positions in the author list while those whose surnames begin with later alphabets perpetually occupy the last position, irrespectively of their contributions, leaving them deeply unsatisfied. To keep younger researchers motivated, some lead researchers or team leaders choose the last position on the author list so that the younger ones occupy the initial positions. In 2012, Harvard University and the Welcome Trust convened a workshop to find a satisfactory solution to the issue of author name order for multi-authored papers. With contributions from research funders, editors, researchers as well as publishers, a *Contributor Roles Taxonomy (CRediT)*, defined in Table 4.1 (Brand et al. 2015), was agreed and it is now adopted by many publishers. The *CRediT Author Statement* states who did what, thereby recognizing individual authors' contribution(s) rather than their position on the author list. Additionally, Table 4.1 can be used as a yard stick in deciding who should be or should not be a co-author (*i.e.*, be on the author list). So, anyone who has performed at least one of the roles in Table 4.1 is eligible to be a co-author. A typical CRediT statement of a paper is shown in Example 4.1. This paper is from my postdoctoral research work at the University of Hull, UK.

Example 4.1 *A CRediT Author Statement of a Paper*

Andrew T. Tyowua: Methodology, Investigation, Visualization, Writing-original draft.
Bernard P. Binks: Funding acquisition, Conceptualization, Supervision, Resources, Validation, Writing-review & editing.
            From Tyowua and Binks (2020, *J. Colloid Interface Sci.* **561**, 127–135)

Based on the CRediT Author Statement in Example 4.1, Tyowua was responsible for the methodology, investigation, visualization and writing-original draft (see definitions in Table 4.1), while Binks was responsible for funding acquisition, conceptualization, supervision, resources, validation, writing-review and editing. Unlike in this paper, where the two contributors are assigned multiple roles, a role may also be assigned to multiple contributors. In this case, the degree of contribution may be optionally specified as "lead", "equal" or "supporting" (Brand et al. 2015). Lastly, the author list is accompanied by institutional or industrial addresses as well as the email address(es) of the corresponding author(s). The corresponding author(s) is/are responsible for all correspondence relating to the paper even after publication. Many journals automatically consider the author submitting the paper as the corresponding author, but another author can also be delegated. If an author has moved to another institution, the institution where the work was done is normally taken as the institutional address while the new institution is considered an affiliated institution and is indicated as *Current Affiliation*.

Activity 4.1 *Author List and CRediT Statement*

   a. Examine the author list of at least five papers of your choice. Are the lead (corresponding) authors also the first authors?
   b. Are the various contributions of the authors specified in the CrediT statement of the papers? Are the contributions of the first author more significant than those of the other authors?

TABLE 4.1
CRediT – Contributor Roles Taxonomy

Term	Definition
Conceptualization	Ideas, formulation or evolution of overarching research goals and aims
Methodology	Development or design of methodology, creation of models
Software	Programming, software development, designing computer programs, implementation of the computer code and supporting algorithms, testing of existing code components
Validation	Verification, whether as a part of the activity or separate, of the overall replication/reproducibility of results/experiments and other research outputs
Formal analysis	Application of statistical, mathematical, computational or other formal techniques to analyze or synthesize study data
Investigation	Conducting a research and investigation process, specifically performing the experiments, or data/evidence collection
Resources	Provision of study materials, reagents, materials, patients, laboratory samples, animals, instrumentation, computing resources or other analysis tools
Data curation	Management activities to annotate (produce metadata), scrub data and maintain research data (including software code, where it is necessary for interpreting the data itself) for initial use and later reuse
Writing – original draft	Preparation, creation and/or presentation of the published work, specifically writing the initial draft (including substantive translation)
Writing – review and editing	Preparation, creation and/or presentation of the published work by those from the original research group specifically critical review, commentary – including pre- or post-publication stages or revision
Visualization	Preparation, creation and/or presentation of the published work, specifically visualization/data presentation
Supervision	Oversight and leadership responsibility for the research activity planning and execution, including mentorship external to the core team
Project administration	Management and coordination responsibility for the research activity planning and execution
Funding acquisition	Acquisition of the financial support for the project leading to this publication

4.2 COMPETING INTERESTS

A competing interest (or conflict of interest) exists when the interpretation or presentation of results is likely to be biased due to the authors' personal or financial relationship with other individuals, organizations or funders or the authors themselves disagree with the interpretation of the findings reported. Authors are expected to declare both financial and non-financial competing interests with respect to the paper. Even if not declared, conflict of interests may become public in the future, *i.e.*, after publication of the paper. Your paper may be retracted from the journal or an Erratum will be required if an undisclosed competing interest becomes known after publication of the paper. Contrarily, the paper may be rejected or the peer-review process may be delayed if an undisclosed competing interest is identified by the editor or the reviewers. Disclosing competing interests does not prevent a paper from being published but helps with the editorial and publication processes. Examples of financial competing interests are

- receiving any form of payment from an organization that may benefit or become disadvantaged by the publication of the paper,
- holding shares or stocks in an organization that may benefit or become disadvantaged by the publication of the paper (read Booth 1988 for a real-life example),
- applying for or holding patents on (similar) contents of the paper and
- receiving any form of payment from an organization holding or applying for patents on the content of the paper.

Examples of non-financial competing interests are

- personal, political or academic gains or losses that may accrue from the publication of the paper,
- disputes among the authors themselves and
- ideological, social or religious stance relating to the content of the paper.

See Irland (2007) for more examples of conflict of interests albeit in the natural resources and environmental management. The conflict of interest section of two papers is given in Example 4.2. The authors of the first paper have no conflict of interests while those of the second paper have several conflicts of interest with respect to the published paper.

Example 4.2 *Conflict of Interest Section of Two Papers. No Conflict of Interest Exists in (a) while Conflicts of Interest Exist in (b)*

a. "The authors declare that they have no known competing financial interests or personal relationships that could have appeared to influence the work reported in this paper". From Tyowua and Ezekwuaku (2021, *Colloids Surf. A* **629**,127386).

b. "Jaime A. Teixeira da Silva was the former founder and CEO/CSO/EIC of Global Science Books (www.globalscienceboks.info). Radha Holla Bhar is a founding member of the Alliance Against Conflict of Interest (AACI), but her ideas in this paper do not necessarily reflect the positions or ideas of AACI. Apart from this, the authors declare no other conflicts of interest. Charles T. Mehlman is the course director for an orthopaedic surgery board review course (Oakstone Publishing) and performs rare/intermittent expert witness testimony, and he serves on the editorial boards of several journals including *Journal of Orthopaedic Trauma, Journal of Pediatric Orthopaedics, The Spine Journal* and *Journal of Children's Orthopaedics*". From Teixeira da Silva et al. (2019, *Bioethical Inquiry* **16**, 279–298).

Activity 4.2 *Conflict of Interest*

Have you ever had conflict of interest(s) in your research team? How was it resolved?

4.3   ACKNOWLEDGMENT(S) SECTION

In this section, as the name implies, the funding organization(s) and all those who contributed to the success of the work and who are not listed as authors, are acknowledged and thanked. Some journals require that the grant or project number(s) be listed along with the funding organizations as shown in Example 4.3 where the grant or project numbers are boldfaced.

Example 4.3 *An Acknowledgment Section of a Paper with Grant or Project Numbers*

"This work was carried out in part within the Australian Research Council (ARC) Centre of Excellence in Convergent Bio-Nano Science and Technology (**Project CE140100036**). A.Z. gratefully acknowledges PEEF Chief Minister Merit Scholarship (CMMS). E.P. thanks CAREER Award **1955170** from the National Science Foundation for financial support. K.K. gratefully acknowledges the award of a NHMRC-ARC Dementia Research Development Fellowship (**APP1109945**) and an ARC Future Fellowship (**FT190100572**) from the Australian Research Council. Figures 1, 13 and the TOC graphic were created in part with BioRender.com".
From Zia et al. (2020, *ACS Appl. Mater. Interfaces* **12**, 38845–38861).

4.4   REFERENCES

The reference section of scientific papers normally contains a comprehensive list of all bibliographic materials (journal papers, textbooks, conference proceedings, patents) cited in the paper. There are hundreds of reference styles with every journal adopting a preferred one. With the readily availability of referencing software, referencing is easier than before when authors compile them manually. Referencing software typically facilitate retrieval, organization and storage of bibliographic materials

(Nashelsky and Earley 1991). There are numerous referencing software, with varying functionality, available at little or no cost. Some common and famous ones include:

- EndNote (http://www.endnote.com) – a paid offline software with individual and institutional license options.
- RefWorks (http://www.refworks.com) – a paid online software which can be purchased by both individuals and institutions, but the institutional package gives more options and features.
- CiteULike (http://www.citeulike.org) – a free offline referencing software that also acts a social networking platform that enables users search related published papers and further connects those with similar research interests.
- Qiqqa (http://www.qiqqa.com) – a free offline open-source referencing software that allows downloading and annotation of pdf documents in addition to its excellent referencing features.
- Mendeley (http://www.mendeley.com) – an online referencing software that gives a free single-user package with the option of upgrading to more users and shared storage space.
- Bookends (http://www.sonnysoftware.com) – a paid offline individual-based software with a free demo version limited to 50 references.
- Citavi (http://www.citavi.com) – a paid offline referencing software that allows downloading and annotation of pdf documents, team collaboration as well as planning and organization of work. It also has a free version with all the features except for a limited number (100) of references.
- Docear (http://www.docear.org) – a free offline open-source referencing software with excellent features that allows downloading and annotation of pdf documents, organization of literature and discovery of open-access papers.
- Zotero (http://www.zotero.org) – a free, online-based, open-source software with Firefox browser plug-in.

More examples of referencing software can be found in Lorenzetti and Ghali (2013). The onus is on you to decide which referencing software is best for you. One obvious advantage of using a referencing software is that it makes referencing very easy, especially when dealing with an enormous number of bibliographic materials.

FURTHER READING

Luther, J. 2015. Recognizing roles and contributions. Learned Publishing 28 (4):236-237.
- Discusses the various ways through which the contributions of researchers and funders toward a published work are now recognized and how it is going to evolve over time.
Richardson, S.M., F. Bella, V. Mougel, and J.V. Milić. 2021. Scientific writing and publishing for early-career researchers from the perspective of young chemists. *J. Mater. Chem. A* 9 (35):18674-18680.
- Contains useful suggestions for writing the various sections of a scientific research paper, including how to decide authorship.

Schultz, D. 2013. *Eloquent Science: A Practical Guide to Becoming a Better Writer, Speaker, and Atmospheric Scientist*: Springer Science & Business Media.
- Contains suggestions for deciding authorship as well as writing other sections of a scientific research paper.

REFERENCES

Booth, W. 1988. Conflict of interest eyed at Harvard. *Science* 242 (4885):1497–1499.

Brand, A., L. Allen, M. Altman, M. Hlava, and J. Scott. 2015. Beyond authorship: Attribution, contribution, collaboration, and credit. *Learned Publishing* 28 (2):151–155.

Irland, L.C. 2007. "Professional ethics for natural resource and environmental managers: A primer." In.: Yale School of Forestry & Environmental Studies.

Lorenzetti, D.L., and W.A. Ghali. 2013. Reference management software for systematic reviews and meta-analyses: An exploration of usage and usability. *BMC Med. Res. Methodol.* 13 (1):141.

Nashelsky, J., and D. Earley. 1991. Reference management software: Selection and uses. *Libr. Software Rev.* 10 (3):174.

Teixeira da Silva, J.A., J. Dobránszki, R.H. Bhar, and C.T. Mehlman. 2019. Editors should declare conflicts of interest. *Bioethical Inquiry* 16 (2):279–298.

Teodosiu, M. 2019. Scientific writing and publishing with IMRaD. *Ann. For. Res.* 62 (2):201–214.

Tyowua, A.T., and B.P. Binks. 2020. Growing a particle-stabilized aqueous foam. *J. Colloid Interface Sci.* 561: 127–135.

Tyowua, A.T., and A.I. Ezekwuaku. 2021. Overcoming coffee-stain effect by particle suspension marble evaporation. *Colloids Surf. A* 629: 127386.

Zia, A., E. Pentzer, S. Thickett, and K. Kempe. 2020. Advances and opportunities of oil-in-oil emulsions. *ACS Appl. Mater. Interfaces* 12 (35):38845–38861.

5 Textual and Video Abstracts

5.1 FUNCTIONS OF THE TEXTUAL ABSTRACT

The primary function of the textual abstract is to give a concise outline or summary of the paper. Therefore, editors use the textual abstract to decide whether or not a paper is suitable for their journal. Similarly, editors first send textual abstracts to reviewers who read them and decide whether or not they are competent enough to review the paper (Cals and Kotz 2013). The editor sends the entire paper only after a reviewer feels competent and agrees to review the paper. Additionally, in combination with the title, textual abstracts are important for *selection* and *indexing*. For selection, textual abstracts help readers to readily determine whether or not the paper is relevant for their purpose and whether or not to read it entirely. For indexing, textual abstracts help quick retrieval of papers from online database search.

5.2 TYPES OF TEXTUAL ABSTRACTS

There are two types of textual abstracts: *descriptive* and *informative* textual abstracts. Nonetheless, these abstracts may be *structured* (with clearly marked subsections) or *unstructured* (without clearly marked subsections). A descriptive textual abstract gives an outline of the information contained in the paper which may include the *aim* and *scope* of the study as well as the *methodology* but leaves out the results and the conclusions. These subsections are clearly marked out in a structured descriptive textual abstract, but they are not clearly marked out in an unstructured descriptive textual abstract. Descriptive textual abstracts are very short (about 100 words or so). Contrarily, an informative textual abstract summarizes the paper under five sub-headings: *background* information, *aim*, *methods*, *results*, *discussion* and the *conclusion*(s) (Weissberg and Buker 1990). That is, an informative textual abstract is a mini-version of the paper. These five subsections are clearly marked out in a structured informative textual abstract, but they are not clearly marked out in an unstructured informative textual abstract. Informative textual abstracts are relatively longer (150–350 words) than descriptive textual abstracts. A descriptive textual abstract is like a *table of contents* of the paper, while an informative textual abstract gives the *content* of the paper. In descriptive textual abstracts, emphases are placed on the problem and the methodology, while in informative textual abstracts, emphases are placed on all the parts of the paper. The textual abstract of many scientific research papers is informative, but textual abstract formats differ across journals. Some journals demand descriptive textual abstracts, while others demand informative textual abstracts with a clear indication of the background, methodology, results and conclusions (*i.e.*, structured). Always consult the *Guide for Authors* of your target

DOI: 10.1201/9781003186748-5

journal for the appropriate format. Comparatively, descriptive textual abstracts are common with uncompleted projects like grant proposals while informative textual abstracts are common with completed projects like full-length scientific research papers. Informative textual abstracts are not only useful in scientific papers involving experiments, but they can also be applied to other forms of writings like review papers (Hartley 2008). Informative textual abstracts of review papers are normally written under three sub-headings (Hartley 2008):

- Background
- Aim
- Conclusions (and recommendations in some cases)

However, it is common to find review papers without the concluding sub-heading like in Example 5.1. Finally, unlike the title (Chapter 3) that conveys the essence of the paper in a single sentence, the textual abstract conveys the essence in the form of a summary using multiple sentences.

Example 5.1 *An Informative Textual Abstract of a Review Paper*

Title of Paper: Honey: Single food stuff comprises many drugs

Aim of Paper: To review physicochemical properties, traditional use of honey as medicine and mechanism of action of honey in the light of modern scientific medicinal knowledge

Abstract

Background: "Honey is a natural food item produced by honey bees. Ancient civilizations considered honey as a God-gifted prestigious product. Therefore, a huge literature is available regarding honey importance in almost all religions. Physically, honey is a viscous and jelly material having no specific color. Chemically, honey is a complex blend of many organic and inorganic compounds such as sugars, proteins, organic acids, pigments, minerals and many other elements. Honey use as a therapeutic agent is as old as human civilization itself. Prior to the appearance of present-day drugs, honey was conventionally used for treating many diseases. At this instant, the modern research has proven the medicinal importance of honey. It has broad spectrum anti-biotic, anti-viral and anti-fungal activities. Honey prevents and kills microbes through different mechanisms such as elevated pH and enzyme activities. Till now, no synthetic compound that works as anti-bacterial, anti-viral and anti-fungal drugs has been reported in honey yet it works against bacteria, viruses and fungi while no anti-protozoal activity has been reported. Potent anti-oxidant, anti-inflammatory and anti-cancerous activities of honey have been reported. Honey is not only significant as anti-inflammatory drug that relieves inflammation but also protect liver by degenerative effects of synthetic anti-inflammatory drugs. *Aim*: This article reviews physicochemical properties, traditional use of honey as medicine and mechanism of action of honey in the light of modern scientific medicinal knowledge".

From Khan et al. (2018, *Saudi J. Biol. Sci.* **25**, 320–325)

Activity 5.1 *Functions and Differences between Descriptive and Informative Textual Abstracts*

a. What are the functions of a textual abstract?
b. What are the differences between a descriptive and an informative textual abstract?

5.3 WRITING THE TEXTUAL ABSTRACT

When writing the textual abstract, be sure that it contains a summary of the following:

a. Descriptive textual abstracts
   - Problem (*research gap identified*) or aim (*research gap addressed in the study*).
   - Method (*how the problem was solved*).
b. Informative textual abstracts
   - Background information in two sentences: (i) an introductory statement that is understandable by both expert and nonexpert audience and (ii) an additional statement understandable by experts in the area.
   - Aim (*research gap addressed in the study*) in a single simple sentence.
   - Methodology (*how the study was done*) in a maximum of three sentences.
   - Results (*what was found from the study*) in a maximum of five sentences.
   - Conclusions (*implications and/or applications of the results*) in a maximum of three sentences. Some recommendations for future research direction can also be added.

Additionally, the textual abstract should be able to stand alone. That is, it should be understood without reference to the paper or other sources because it is always made available online free of charge without other parts of the paper, making it widely accessible. The textual abstract is the most read part of the paper – many people stop at it and do not read the entire paper. In order for the textual abstract to stand alone, avoid the following:

- Abbreviations.
- Symbols.
- Mathematical formulae.
- References.

Also, for the purpose of indexing, include keywords researchers would likely use while searching for papers in the area. Examples 5.2 and 5.3 are, respectively, typical descriptive and informative textual abstracts. The key finding is also included in the descriptive textual abstract, but this is optional. The informative textual abstract contains a concise summary of the background of the study (one sentence), the methodology (two sentences), the results (five sentences) and the conclusions (three sentences). Finally, the abstract is written in the *present* or *past tense* depending on the subsection (see Chapter 14 for the use of tenses in different parts of a research paper).

Example 5.2 *An Example of a Descriptive Textual Abstract from a Published Research Paper*

Title of Paper: Fired clay bricks using agricultural biomass wastes: Study and characterization

Aim of Paper: To investigate the effects of the incorporation of renewable pore forming agents (wheat straw, sunflower seed cake and olive stone flour) on the properties of fired bricks

Abstract

"Aim: The main objective of this study is to investigate the effects of the incorporation of renewable pore forming agents on the properties of fired bricks. *Methods*: Different additives have been studied (wheat straw, sunflower seed cake and olive stone flour) at different grinding and incorporation rate. Physical properties such as linear shrinkage, loss on ignition, bulk porosity, water absorption and bulk density have been measured. Mechanical and thermal performances have also been characterized. *Key finding*: The incorporation of 4wt% of sunflower seed cake, with the lowest grinding, leads to the best compromise between mechanical and thermal results compared to the reference brick".

From Bories et al. (2015, *Constr. Build Mater.* **91**, 158–163)

Example 5.3 *An Example of an Informative Textual Abstract from a Published Research Paper*

Title of Paper: Tobacco, alcohol and cannabis use during pregnancy: Clustering of risks

Aim of Paper: This paper examines self-reported and concurrent use of tobacco, alcohol and cannabis among pregnant indigenous (Aboriginal and Torres Strait Islander) women and compares characteristics of women by the number of current substances reported.

Abstract

"Background: Antenatal substance use poses significant risks to the unborn child. We examined use of tobacco, alcohol and cannabis among pregnant Aboriginal and Torres Strait Islander women; and compared characteristics of women by the number of substances reported.

Methods: A cross-sectional survey with 257 pregnant indigenous women attending antenatal services in two states of Australia. Women self-reported tobacco, alcohol and cannabis use (current use, ever use, changes during pregnancy); age of initiation of each substance; demographic and obstetric characteristics.

Results: Nearly half the women (120; 47% (95% confidence interval: 40%, 53%)) reported no current substance use; 119 reported current tobacco (46%; 95% confidence interval: 40%, 53%), 53 (21%; 95% confidence interval: 16%, 26%) current alcohol and 38 (15%; 95% confidence interval: 11%, 20%) current cannabis use. Among 148 women smoking tobacco at the beginning of pregnancy, 29 (20%; 95% confidence interval: 14%, 27%) reported quitting; with 80 of 133 (60%; 95% confidence interval: 51%, 69%) women quitting alcohol and 25 of 63 (40%; 95% confidence interval: 28%, 53%) women quitting cannabis. Among 137 women reporting current substance use, 77 (56%; 95% confidence interval: 47%, 65%) reported one and 60 (44%; 95% confidence interval: 35%, 53%) reported two or three. Women using any one substance were significantly more likely to also use others. Factors independently associated with current use of multiple substances were years of schooling and age of initiating tobacco.

Conclusions: While many women discontinue substance use when becoming pregnant, there is clustering of risk among a small group of disadvantaged women. Programs should address risks holistically within the social realities of women's lives rather than focusing on individual tobacco smoking. Preventing uptake of substance use is critical".

From Passey et al. (2014, *Drug Alcohol Depend.* **134**, 44–50)

Depending on the journal, a textual abstract may be structured or unstructured. Even when structured, one or more of the various subsections (*i.e.*, background, aim, methodology, conclusion, etc.) is/are often omitted like in Example 5.4 where the conclusion is omitted.

Example 5.4 *An Example of a Structured Informative Textual Abstract without a Conclusion*

Title of Paper: Synergistic effects of barley, oat and legume material on physicochemical and glycemic properties of extruded cereal breakfast products

Aim of Paper: To investigate the effects of combining barley, oat and legume material into a novel cereal breakfast in relation to both physicochemical and glycemic properties

Abstract

"*Background*: Cereal products are a main source of carbohydrates in the human diet. Barley and oat grains are considered as functional ingredients that can manipulate postprandial glycemic load while legume material provides high fiber and protein components to food. *Aim*: This research reports on hot extrusion (barrel temperature: 81.9–103.6 °C, die temperature: 57.3–77.3 °C and screw speed was around 200 rpm) to develop a barley-oat breakfast cereal which were supplemented with 10% green or yellow pea. *Methodology*: The impact of barley-oat ratio and additional pea on physicochemical properties of extruded cereal breakfast was determined. *Results*: Generally, oat positively affected the hardness of the extrudates, and the hardest extrudates were obtained when the barley-oat ratio was 50:50%. The presence of green and yellow pea increased greenness, redness, chroma and browning index of the samples. The results of *in-vitro* starch digestion test indicated that increase of oat ratio significantly decreased rapidly digested carbohydrates and starch digestibility AUC (area under the curve) ($P < 0.05$). However, the presence of peas did not significantly affect the starch digestibility AUC value".

From Brennan et al. (2016, *J. Food Process. Pres.* **40**, 405–413)

Compared with unstructured abstracts, structured abstracts are (i) easier to read, (ii) easier to search, (iii) more useful in peer-review especially for conferences, and (iv) more acceptable by authors and by readers (Hartley 2004). However, unstructured abstracts are more coherent than structured abstracts because the sentences in the various subsections of the latter do not normally cohere with one another unlike those in the former that cohere with one another.

Activity 5.2 *Analyzing Textual Abstracts into Various Subsections and Identification of Tenses Used*

Analyze the following textual abstracts into various subsections in line with Section 5.3, indicating where the present or past tense is used.

a. **Title of paper:** Suppression of the coffee-ring effect by sugar-assisted depinning of contact line

   Aim: To investigate the effects of sugar on the deposition pattern of suspended particles

   Abstract: "Inkjet printing is of growing interest due to the attractive technologies for surface patterning. During the printing process, the solutes are transported to the droplet periphery and form a ring-like deposit, which disturbs the fabrication of high-resolution patterns. Thus, controlling the uniformity of particle coating is crucial in the advanced and extensive applications. Here, we find that sweet coffee drops above a threshold sugar concentration leave uniform rather than the ring-like pattern. The evaporative deposit changes from a ring-like pattern to a uniform pattern with an increase in sugar concentration. We moreover observe the particle movements near the contact line during the evaporation, suggesting that the sugar is precipitated from the droplet edge because of the highest evaporation and it causes the depinning of the contact line. By analyzing the following dynamics of the depinning contact line and flow fields and observing the internal structure of the deposit with a focused ion beam scanning electron microscopy system, we conclude that the depinned contact line recedes due to the solidification of sugar solution without any slip motion while suppressing the capillary flow and homogeneously fixing suspended particles, leading to the uniform coating. Our findings show that suppressing the coffee-ring effect by adding sugar is a cost-effective, easy and nontoxic strategy for improving the pattern resolution".

   From Shimobayashi et al. (2018, *Sci. Rep.* **8**(2) 17769)

b. **Title of paper:** Fermentation and germination improve nutritional value of cereals and legumes through activation of endogenous enzymes

   Aim: To provide a review of how fermentation and germination influence nutrient content and availability

   Abstract: "Cereals and legumes are outstanding sources of macronutrients, micronutrients, phytochemicals, as well as antinutritional factors. These components present a complex system enabling interactions with different components within food matrices. The interactions result in insoluble complexes with reduced bioaccessibility of nutrients through binding and entrapment thereby limiting their release from food matrices. The interactions of nutrients with antinutritional factors are the main factor hindering nutrients release. Trypsin inhibitors and phytates inherent in cereals and legumes reduce protein digestibility and mineral release, respectively. Interaction of phytates and phenolic compounds with minerals is significant in cereals and legumes. Fermentation and germination are commonly used to disrupt these interactions and make nutrients and phytochemicals free and accessible to digestive enzymes. This paper presents a review on traditional fermentation and germination processes as a means to address myriad interactions through activation of endogenous enzymes such as α-amylase, pullulanase, phytase, and other glucosidases. These enzymes degrade antinutritional factors and break down complex mac-ronutrients to their simple and more digestible forms".

From Nkhata et al. (2018, *Food Sci. Nutr.* **6**, 2446–2458)

5.4 VIDEO ABSTRACTS

A video abstract is an audio-visual form of the textual abstract. Video abstracts are relatively short (< 10 min), and they are akin to the sneak preview of a movie. As a preview, they lure the reader to reading the entire paper for details of the research. In fact, a strong positive correlation has been observed between video abstract views of papers and the number of reads for the same papers (Spicer 2014). The *Journal of Visualized Experiments* (*JOVE*) has been publishing professional peer-reviewed experimental videos since 2007 and thus can be considered the pioneer of video abstracts. Examples of video abstracts can be found on the *JOVE* website and websites of other journals like *Cell Press* and *Biotechnology and Bioengineering* that publish video abstracts. Similar to video abstracts are video bytes which give a one-minute overview of a research and its impact on society. Video bytes are geared toward nonexperts of the subject area and are published on social media. Video bytes are made from images generated from the research, and they are produced with or without background music.

5.4.1 TYPES OF VIDEO ABSTRACTS

There are eight different types/forms of video abstracts (Slattery 2020):

i. *Do-it-yourself whiteboard explainer video abstracts.* The research is explained on a whiteboard, blackboard or white paper by drawing sketches and illustrations using a pen or marker. Requiring a webcam or camera and an audio recorder, these video abstracts are the simplest to make, provided the author is good at drawing. However, care must be taken to ensure that nothing obscures the drawings and no background noise interferes with the audio recording. Examples of these abstracts are the *Cell Press* video abstracts "Herd immunity: Understanding COVID-19" and "Have you flossed your gut mucosa today?" on YouTube.

ii. *Animated whiteboard video abstracts.* The research is explained on a whiteboard by drawing sketches and illustrations using an appropriate drawing software. The drawings here are more professional and visually appealing and can be produced more readily than those in (i). Examples include the video abstracts "Species on the move: impacts on ecosystems and human well-being" published by the journal *Science* and "Flow and fishes of the Murray river" produced by *Animate Your Science TV*.

iii. *Motion graphics animation video abstracts.* Unlike in (i) and (ii) where the drawings are motionless, here the drawings are animated with a characteristic movement, *e.g.*, the video abstract "Witnessing causal nonseparability" and "Parallel planning teaches self-driving cars to respond quickly to emergencies" published respectively by the journals *New Journal of Physics* and *Institute of Electrical and Electronics Engineers* on YouTube. Therefore, these video abstracts are more dynamic than those in (i) and (ii).

iv. *Narrated slides video abstracts.* These video abstracts are very similar to oral presentations discussed in Chapter 21. Therefore, the suggestions given in Chapter 21 with respect to oral presentation apply here too. The slides are prepared using an appropriate software, and they are recorded using a screen recorder as the author explains the research, *e.g.*, the *Cell Press* video abstracts "Origin of cell-size homeostasis in bacteria" and "Emerging optical nanoscopy techniques" on YouTube. These video abstracts are straightforward to make, especially for those good at giving oral presentation.

v. *Talking head video abstracts.* Contains a portrait video footage of the author explaining the research. In addition to the research, the video helps put a face to the name of the author. These videos are a little boring because they lack visuals. It is a great idea to alternate between the video footage of the author and other footages (*e.g.*, slides, photographs, sketches, *etc.*). Examples of talking head video abstracts are the *Cell Press* video abstracts "Developmental Cell Levin/Yanai" and "A developmental clock within the brain" on YouTube.

vi. *Experimental footage video abstracts.* The research is narrated verbally in combination with a video footage of the experimental procedure, sample collection, important finding(s), *etc.* Examples of these abstracts are the *Cell Press* video abstract "Natural selection and spatial cognition" and the *Canadian Journal of Fisheries and Aquatic Sciences* video abstract "A spatial capture–recapture model to estimate fish survival and location from linear continuous monitoring arrays" on YouTube.

vii. *Text overlay video abstracts.* The research is narrated by short text (< 15 words per instance) overlaid on a footage of the experimental procedure, sample collection, important finding(s), images, *etc.* Such videos are normally with or without background music. Examples include the *Wiley* video abstracts "The deep physiological connections that form among choir singers" and "The coldest places on earth" on YouTube.

viii. *Multiple-styled video abstracts.* The research is explained using a combination of video abstract styles, *e.g.*, a talking head and narrated slides like the *Cell Press* video abstract "Amygdala reward neurons form and store fear extinction memory" on YouTube.

Endeavor to watch the exemplary videos given here as they will help you understand the differences between the various types of video abstracts.

Activity 5.3 *Categorizing Video Abstracts into Various Types*

Watch the following Cell Press video abstracts on YouTube and categorize them into various types in line with Section 5.4.1.

a. Cancer mystery goes up in smoke
b. Multiple sclerosis genetics hits a nerve
c. Gut microbes and heart disease
d. Natural killer cells: An anatomical view
e. Video abstract – Salt-induced edible non-spherical Pickering emulsions droplets

5.4.2   BENEFITS OF VIDEO ABSTRACTS

Video abstracts have several benefits compared with their textual counterpart. For example, compared with textual abstracts where readers are expected to read and make sense out of it, no reading is required with video abstracts, except for videos that contain either subtitles or overlaid text. By simply clicking the play button of the video, the reader receives first-hand and concise information about the research. Also, video abstracts give authors the opportunity to dramatize complex concepts in the most visually appealing ways. Furthermore, video abstracts enable communication of scientific findings to a much broader audience, and they are ideal for sharing on social media platforms. Therefore, scientific journals and publishers are now increasingly using video abstracts to effectively communicate science and to also market their publications to a broader audience, thus increasing the visibility and the impact of their publications. Finally, video abstracts have the potential to increase funding and job opportunities for authors. The downside, however, is that making a video abstract requires very particular sets of skills, but the good news is that professional graphic designers can help make your video abstract at an affordable price. Such designers can be found both online and offline.

5.4.3   CONTENTS OF VIDEO ABSTRACTS

Generally, video abstracts explain why and how the research was done and then highlight the most important finding. Specifically, video abstracts contain the following:

- *Introduction* – A description of problems or questions tackled in the paper. The videos state the research gap identified in the literature and the importance of filling the gap.
- *Methodology* and *results* – A description of the experiments used in tackling the research problem or question and the results or data obtained. That is, an explanation of the approach used in closing the research gap and the corresponding observations. This is where the power of a video abstract is felt most. Therefore, irrespective of the complexity of the science involved, use clear and precise illustrations that will be understood by both experts and nonexperts of the subject area.

- *Conclusions* – Highlights of the most important conclusions and the take-home message.
- *Invitation* – An invitation to read the entire paper for full details. This is accompanied by a downloadable link of the published paper.

However, this is not a strait jacket. The contents, length and technical specifications, *e.g.*, file format (mp4, mov, *etc.*), file size, aspect ratio, frame rate and type, of video abstracts vary from journal-to-journal. Always remember to consult the *Guide for Authors* for video abstracts guidelines of the journal.

5.4.4 MATERIALS FOR MAKING A VIDEO ABSTRACT

If you have been making teaching videos, making a video abstract will be much easier because you would have had most of the materials and skillsets needed. However, the materials needed for making a video abstract are dependent on the video type. Therefore, depending on the video type, you will probably need all or some of these materials:

- A computer to house your software applications, prepare scripts or text, prepare slides, *etc.*
- Animation software – This will enable you to explain your research using animation. There are several paid and free animation software to choose from, *e.g.*, adobe animate and blender are paid and free animation software, respectively. Microsoft PowerPoint can also be used to animate research.
- Audio/voice recorder – This will enable you to record your voice while explaining the research. There are also paid and free voice recording software, *e.g.*, Adobe audition and Audacity are, respectively, paid and free voice recording and editing software applications. Other free voice recording and editing software include DarkAudacity, Shotcut and Ocenaudio. Voice recording can also be done with Microsoft PowerPoint. These software applications work with the in-built microphone of your computer. Therefore, if the microphone does not serve your purpose, get an external one.
- Webcam or camera – This will enable you to record a footage of yourself as you explain the research.
- Screen recording software – This will enable you to record your animation and video clips directly from your computer screen. Some free screen recording software include OBS, Bandicam and Vimeo. Microsoft PowerPoint can also be used for screen recording.
- Video editing software – This will enable you to edit your video to an acceptable quality. Davinci Resolve is a popular free video editing software with advanced features.

There are several alternatives for all the software applications mentioned here. Be sure to look around for them, if necessary.

5.4.5 Making a Video Abstract

The following tips will help you make a great video abstract:

i. Decide the type of video abstract you want.

ii. Acquire the materials needed for producing the video abstract you have decided. Learn any software applications needed thoroughly.

iii. Think about the script and contents of the video and prepare them accordingly. The script should take the format given in Section 5.4.3 or any order of your choice. The contents could be images, illustrations, animations, overlaid text, *etc.*

iv. If your footage is going to be part of the video, do not read the script on the webcam or camera as this will make your video unnatural.

v. If you are going to record slide(s), use font type and size that will be legible to users of all devices (*i.e.*, phones, computers, tablets, *etc.*). In addition, such slides should not be loaded with text. In fact, use less text and more animations or illustrations. Also, do not use raw data and complex figures (*i.e.*, graphs and charts). In fact, the guide given in Chapters 20 and 21 for preparing posters and oral presentation figures also applies here.

vi. Explain the research in clear and plain language that will be understood by both experts and nonexperts of the subject area.

vii. Practice and practice before making your video, especially the script. This will save you from frustration as well as editing time.

viii. Record in a well-lit and noiseless room. No amount of editing will make a poor video great.

ix. Adhere strictly to the video abstract guidelines of the journal.

x. If the video length is not prescribed by the journal, keep your video to a maximum of 5 min.

xi. To optimize the audio and video quality, consider recording them separately. This will also allow you to edit them separately.

FURTHER READING

Katz, M.J. 2009. *From Research to Manuscript: A Guide to Scientific Writing*: Springer Science & Business Media.
- Contains a comprehensive guide for writing the abstract and other sections of a research paper with excellent examples drawn from papers published by reputable scientific journals.

Parija, S.C., and V. Kate, eds. 2017. *Writing and Publishing a Scientific Research Paper*: Springer, Singapore.
- Contains valuable suggestions for writing all the sections of a scientific research paper. Specifically, Chapter 4 contains the etymology of the word "abstract", its definition and suggestions for writing the abstract section of a scientific research paper.

Slattery, F. 2020. "8 ways to make a video abstract - with examples." https://www.animateyour.science/post/8-ways-to-make-a-video-abstract
- Discusses the different types of video abstracts and gives useful suggestions for creating each abstract type.

Weissberg, R., and S. Buker. 1990. *Writing Up Research*: Prentice Hall Englewood Cliffs, NJ.

- An excellent guide for writing up all the sections of a scientific research paper. Suggestions for writing and reducing the length of a textual abstract are also given.

REFERENCES

Bories, C., L. Aouba, E. Vedrenne, and G. Vilarem. 2015. Fired clay bricks using agricultural biomass wastes: Study and characterization. *Constr. Build. Mater.* 91:158–163.

Brennan, M.A., T. Lan, and C.S. Brennan. 2016. Synergistic effects of barley, oat and legume material on physicochemical and glycemic properties of extruded cereal breakfast products. *J. Food Process. Pres.* 40 (3):405–413.

Cals, J.W., and D. Kotz. 2013. Effective writing and publishing scientific papers, part II: Title and abstract. *J. Clin. Epidemiol.* 66 (6):585.

Hartley, J. 2004. Current findings from research on structured abstracts. *J. Med. Libr. Assoc.* 92 (3):368.

Hartley, J. 2008. *Academic Writing and Publishing: A Practical Handbook*: Routledge, New York.

Khan, S.U., S.I. Anjum, K. Rahman, M.J. Ansari, W.U. Khan, S. Kamal, B. Khattak, A. Muhammad, and H.U. Khan. 2018. Honey: Single food stuff comprises many drugs. *Saudi J. Biol. Sci.* 25 (2):320–325.

Nkhata, S.G., E. Ayua, E.H. Kamau, and J.-B. Shingiro. 2018. Fermentation and germination improve nutritional value of cereals and legumes through activation of endogenous enzymes. *Food Sci. Nutr.* 6 (8):2446–2458.

Passey, M.E., R.W. Sanson-Fisher, C.A. D'Este, and J.M. Stirling. 2014. Tobacco, alcohol and cannabis use during pregnancy: Clustering of risks. *Drug Alcohol Depend.* 134:44–50.

Shimobayashi, S.F., M. Tsudome, and T. Kurimura. 2018. Suppression of the coffee-ring effect by sugar-assisted depinning of contact line. *Sci. Rep.* 8 (1):17769.

Slattery, F. 2020. "8 ways to make a video abstract - with examples." https://www.animateyour. science/post/8-ways-to-make-a-video-abstract.

Spicer, S. 2014. Exploring video abstracts in science journals: An overview and case study. *JLSC* 2 (2): eP1110.

Weissberg, R., and S. Buker. 1990. *Writing Up Research*: Prentice Hall Englewood Cliffs, NJ.

6 Keywords and Phrases

6.1 FUNCTIONS OF KEYWORDS AND PHRASES

Keywords and phrases (also known as *subject terms*) are respectively words and group of words taken from either the textual abstract or the text of a paper that describe the content or essence of the paper. This can be compared with the title that describes the essence of the paper using a single sentence and the textual abstract that describes the essence of the paper using multiple sentences. Therefore, the main function of keywords and/or phrases is to complement the title, the abstract and the author information (specifically, author name(s)) by enabling easy retrieval of the paper from bibliographic searches, which ultimately increases scholarly visibility (Gbur and Trumbo 1995). Keywords also (Hartley 2008):

- Help readers to quickly decide whether the paper is relevant for their purpose or not.
- Enable the paper to be grouped together with other related papers.
- Help in tracking the growth of the subject area.

6.2 SELECTING KEYWORDS AND PHRASES

Many scientific journals now require authors to select three to six keywords and/or phrases that best describe their work. However, there is little information on how to select effective keywords and phrases. A rough guide for selecting keywords and phrases is to ask yourself what words you will input into a search engine if you were to search for information on the topic of your manuscript (Mack 2012). You may come up with two to four words or phrases. Perform a search with them; if they do not bring too many results or off-scope articles, then they are effective keywords and phrases for your manuscript. Otherwise, choose alternative words and try the search again until you find words and phrases with fewer search results and off-scope papers. Some medical journals require authors to select keywords from terminologies indexed by established databases like Medical Subject Headings (MeSH) list of Index Medicus (Bahadoran et al. 2020), but the same approach can be used to select effective keywords from the list. Keywords are sometimes supplied by editors or referees as well as computer programs at the proof stage (Hartley and Kostoff 2003). Additionally, Gbur and Trumbo (1995) have given ten suggestions for selecting effective keywords and phrases which can guide authors, editors as well as referees in selecting keywords. These suggestions are summarized as follows (Gbur and Trumbo 1995):

DOI: 10.1201/9781003186748-6

    i. Use simple, specific noun clauses, *e.g.*, use *variance estimate* instead of *estimate of variance.*

    ii. Avoid words and phrases that are too common to minimize the search results to a manageable number.

    iii. Avoid words and phrases already in the title because the title and the keywords are normally indexed. This is contrary to Mack (2012) that encouraged the use of keywords and phrases in the title and abstract, in agreement with Example 6.1.

    iv. Where possible, avoid the use of prepositions like *in* and *of*, *e.g.*, use *data quality* instead of *quality of data.*

    v. Avoid acronyms because they are not likely to be used in search by those who are new in the area.

    vi. Avoid Greek letters and mathematical symbols because it is impossible to use them in computer-based searches.

    vii. Do not include names of people unless they are part of an established terminology, *e.g.*, *"Poisson" distribution.*

    viii. Include important mathematical or computer techniques, *e.g.*, *generating function*, used to obtain results, and a statistical philosophy or approach, *e.g.*, *maximum likelihood* or *Bayes' theory.*

    ix. Where appropriate, include areas of applications.

    x. Be sure to include alternative or inclusive terminologies – where a concept is or has been known by many terminologies, use a keyword that will enable a search across a time-span.

Finally, machine learning has been proposed to support keyword selection (Wu et al. 2007).

Example 6.1 *Analysis of Paper Title and Abstract for the Use of Keywords*

The keywords and phrases of the following paper are "indigenous", "Aboriginal and Torres Strait Islander", "prenatal care", "prevention", "Australia" and "tobacco". Check whether or not the paper title and abstract contain these keywords and phrases.

Title of Paper: Tobacco, alcohol and cannabis use during pregnancy: Clustering of risks

Abstract

"Background: **Antenatal** substance use poses significant risks to the unborn child. We examined use of **tobacco**, alcohol and cannabis among pregnant **Aboriginal and Torres Strait Islander women**; and compared characteristics of women by the number of substances reported.

Methods: A cross-sectional survey with 257 pregnant indigenous women attending **antenatal services** in two states of **Australia**. Women self-reported **tobacco**, alcohol and cannabis use (current use, ever use, changes during pregnancy); age of initiation of each substance; demographic and obstetric characteristics.

Results: Nearly half the women (120; 47% (95% confidence interval: 40%, 53%)) reported no current substance use; 119 reported current **tobacco** (46%; 95%

confidence interval: 40%, 53%), 53 (21%; 95% confidence interval: 16%, 26%) current alcohol and 38 (15%; 95% confidence interval: 11%, 20%) current cannabis use. Among 148 women smoking **tobacco** at the beginning of pregnancy, 29 (20%; 95% confidence interval: 14%, 27%) reported quitting; with 80 of 133 (60%; 95% confidence interval: 51%, 69%) women quitting alcohol and 25 of 63 (40%; 95% confidence interval: 28%, 53%) women quitting cannabis. Among 137 women reporting current substance use, 77 (56%; 95% confidence interval: 47%, 65%) reported one and 60 (44%; 95% confidence interval: 35%, 53%) reported two or three. Women using any one substance were significantly more likely to also use others. Factors independently associated with current use of multiple substances were years of schooling and age of initiating **tobacco**.

Conclusions: While many women discontinue substance use when becoming pregnant, there is clustering of risk among a small group of disadvantaged women. Programs should address risks holistically within the social realities of women's lives rather than focusing on individual **tobacco** smoking. **Preventing** uptake of substance use is critical".

From Passey et al. (2014, *Drug Alcohol Depend.* **134**, 44–50)

All the four keywords and three key phrases in Example 6.1 have been used in both the title and the abstract, either exactly or in a slightly different form, e.g., parental care as antenatal services. Specifically, the title contains the keyword "tobacco", while the abstract contains all the keywords and phrases. These words and phrases have been **underlined** to enable quick visualization.

Activity 6.1 *Searching for Keywords and Phrases in the Title and Abstract of Published Scientific Papers*

Analyze at least five paper titles and abstracts of published articles in your subject area for the use of keywords and phrases in either the title or abstract or both. You will notice that many published articles contain a substantial number of listed keywords and phrases in their titles and abstracts.

FURTHER READING

Parija, S.C., and V. Kate, eds. 2017. *Writing and Publishing a Scientific Research Paper*: Springer, Singapore.
- Chapter 4 contains useful suggestions for selecting keywords for a scientific research paper. The book also contains suggestions for writing other sections of the paper.
Schultz, D. 2013. *Eloquent Science: A Practical Guide to Becoming a Better Writer, Speaker, and Atmospheric Scientist*: Springer Science & Business Media.
- Contains suggestions for selecting keywords for a scientific research paper as well as suggestions for writing other sections of the paper.

REFERENCES

Bahadoran, Z., P. Mirmiran, K. Kashfi, and A. Ghasemi. 2020. The Principles of Biomedical Scientific Writing: Abstract and Keywords. *Int. J Endocrinol. Metab.* 18 (1):e100159.
Gbur, E.E., and B.E. Trumbo. 1995. Key Words and Phrases—The Key to Scholarly Visibility and Efficiency in an Information Explosion. *Am. Stat.* 49 (1):29–33.

Hartley, J. 2008. *Academic Writing and Publishing: A Practical Handbook*: Routledge.

Hartley, J., and R.N. Kostoff. 2003. How Useful are 'Key Words' in Scientific Journals? *J. Inf. Sci.* 29 (5):433–438.

Mack, C.A. 2012. How to write a good scientific paper: title, abstract, and keywords. *J. Micro-Nanolith. MEMS.* 11 (2):020101.

Passey, M.E., R.W. Sanson-Fisher, C.A. D'Este, and J.M. Stirling. 2014. Tobacco, alcohol and cannabis use during pregnancy: Clustering of risks. *Drug Alcohol Depend.* 134:44–50.

Wu, C., M. Marchese, J. Jiang, A. Ivanyukovich, and Y. Liang. 2007. Machine learning-based keywords extraction for scientific literature. *J. UCS* 13 (10):1471–1483.

7 Graphical Abstract

7.1 FUNCTIONS OF THE GRAPHICAL ABSTRACT

A graphical abstract is a figure that gives a concise visual description or summary of the key message of a scientific paper. The figure is either a molecule, a reaction mechanism or pathway, a graph, a photograph, a sketch or a combination of one or more of these (with text in some cases). The publication of graphical abstracts began in chemistry journals, namely *Angewandte Chemie, Tetrahedron Letters, Chemical Communications* and the *Journal of American Chemical Society* (Lane et al. 2015). Nonetheless, over 90% of scientific journals now publish graphical abstracts in addition to the textual abstract. Consequently, graphical abstracts have evolved across scientific journals and are now referred to by various names – visual abstract, table of content image or graphic – depending on how the journal views the function of the graphical abstract. The sole function of a graphical abstract is to give a visual description of the main idea (or essence) of the paper at a glance (Pferschy-Wenzig et al. 2016). Furthermore, graphical abstracts (Lane et al. 2015, Pferschy-Wenzig et al. 2016):

- Readily draw readers' attention to papers, thereby increasing readership.
- Are more suitable for sharing on social media than textual abstracts.
- Help readers to readily identify papers that are most relevant for their purpose.
- Enable readers to quickly scan through published volumes of journals online.
- Facilitate browsing and retrieval of papers in online image searches.
- Promote interdisciplinary collaboration or research.

7.2 TYPES OF GRAPHICAL ABSTRACTS

A graphical abstract may contain one or more of the following elements: a molecule, a reaction mechanism or pathway, a graph, a photograph, a sketch or a group of sketches. Lane et al. (2015) have created a four-level descriptive taxonomy for graphical abstracts published in scientific papers which can be summarized as: (i) visual composition, (ii) rhetorical purpose, (iii) modal relationship and (iv) mediated text.

i. **Visual composition:** This description is based on the number of elements used and their interrelationship. Graphical abstracts in this category are further subdivided into singular, stacked, relational and blended visual composition graphical abstracts.

DOI: 10.1201/9781003186748-7

- *Singular visual composition* – Contains a single element, *e.g.*, molecule, graph, photograph or sketch, *e.g.*, Figure 7.1a where the authors reported the fabrication of a flexible aluminum-air battery (Ma et al. 2019).
- *Stacked visual composition* – Contains two or more elements stacked together without any link or relationship between them. These graphical abstracts are difficult to comprehend; therefore, they are uncommon. An example of a stacked visual graphical abstract is shown in Figure 7.1b where the authors showed that water droplets evaporate faster on heated non-wettable vertical substrates than on corresponding horizontal substrates (Qi et al. 2019). Notice that the elements are stacked together without any link or connection between them, making it difficult to understand the key message of the graphical abstract.
- *Relational visual composition* – Contains two or more elements stacked together and connected or linked by arrows. An example is shown in Figure 7.1c where the authors reported that the crack density of bloodstains increases with increase in blood viscosity and decrease in environmental humidity (Choi et al. 2020). The graphical abstract

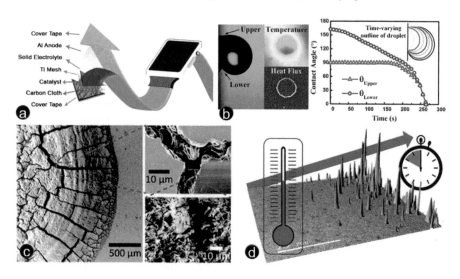

FIGURE 7.1 Examples of different visual composition graphical abstracts. (a) A singular visual graphical abstract containing an image of a digital wrist watch. From Ma et al. (2019, *ACS Appl. Mater. Interfaces* **11**, 1988–1995) (b) A stacked visual graphical abstract containing an image of an evaporating water droplet on a vertical substrate and a graph showing how the upper and the lower contact angles of the droplet change with time. From Qi et al. (2019, *Langmuir* **35**, 17185–17192) (c) A relational visual graphical abstract containing a microscope image of a dried blood sample at low, moderate and high magnifications. From Choi et al. (2020, *Soft Matter* **16**, 5571–5576) (d) A blended visual graphical abstract containing a transition electron microscope image of nanocrystals of osmium superimposed with sketches of a thermometer and a stopwatch, indicating the growth of the nanocrystals with time and increase in temperature. From Pitto-Barry and Barry (2019, *Angew. Chem. Int. Ed.* **58**, 18482–18486).

(Figure 7.1c) is hierarchical in nature, showing a typical bloodstain crack at different magnifications.

- *Blended* (or *hybrid*) *visual composition* – Contains a single image that is blend with two or more superimposed elements. For example, Figure 7.1d where the authors showed that the nucleation and growth of precious metal nanocrystals increase with increasing temperature (Pitto-Barry and Barry 2019).

ii. **Rhetorical purpose:** This description is based on the function of the graphical abstract in the paper. In this case, graphical abstracts are considered as either topical/descriptive or argumentative.

- *Descriptive rhetorical purpose*: Describes or illustrates one or more key concepts but does not relate them or make any claims. For example, Figure 7.2a which illustrates the concept of qualitative analysis of ions from inorganic salt solutions as well as acid-base titration in a liquid marble (*i.e.*, a microliter liquid drop coated with micrometer-sized particles) reported by Tyowua et al. (2020).

- *Argumentative rhetorical purpose*: Makes a claim and provides one or more evidence to support it. An example of an argumentative graphical abstract is shown in Figure 7.2b where the authors reported the creation of a thermoresponsive transparent and impact resistant polymer, poly(glycerol-dodecanoate), with application in smart window fabrication (Zhang et al. 2019). The authors claimed that windows coated with the polymer are (i) transparent when heated and translucent when cooled and (ii) resistant to impact (Zhang et al. 2019). These are the claims illustrated in the graphical abstract (Figure 7.2b). Sometimes, argumentative graphical abstracts may follow the rhetorical moves of textual abstracts – background, carving research gap, creating interest in the work, describing methods, showing results or significance of results or claim(s).

iii. **Modal relationship:** This description is based on the relationship between the graphical and the textual abstracts. On this basis, graphical abstracts are redundant, complementary, supplementary or stage-setting.

- *Redundant modal relationship* – The graphical and the textual abstracts have identical contents, *e.g.*, Figure 7.3 (Hönes et al. 2020) where oil is fused into a gel matrix to create a surface that prevents polydopamine attachment. Through its phenolic hydroxyl groups, polydopamine has the ability to attach to almost all surfaces, including hydrophobic ones. The textual abstract says:

"Polydopamine (PDA) is well-known as the first material-independent adhesive, which firmly attaches to various substances, even hydrophobic materials, through strong coordinative interactions between the phenolic hydroxyl groups of PDA and the substances. In contrast, oil-infused materials such as self-lubricating gels (SLUGs) exhibit excellent antiadhesive properties against viscous liquids, ice/snow, (bio)fouling and so on. In this study, we simply questioned: "What will happen when these two materials

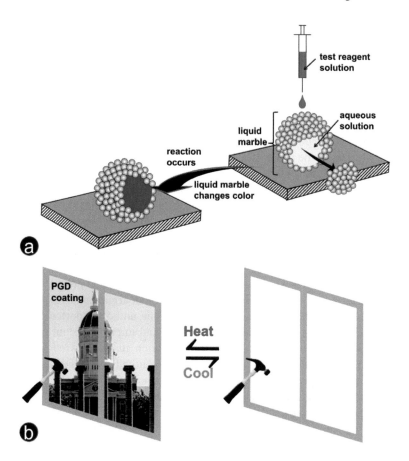

FIGURE 7.2 Examples of different rhetorical purpose graphical abstracts. (a) A topical or descriptive rhetorical purpose graphical abstract represented by a sketch illustrating qualitative inorganic analysis of ions and acid-base titration in an aqueous liquid marble. From Tyowua et al. (2020, *SN Appl. Sci.* **2**, 345) (b) An argumentative rhetorical purpose graphical abstract represented by a sketch showing that a tough hybrid polymer is transparent when heated, but translucent when cooled. From Zhang et al. (2019, *ACS Appl. Mater. Interfaces* **11**, 5393–5400).

with contrary nature meet"? To answer this, we formed a PDA layer on a SLUG surface that exhibits thermoresponsive syneretic properties (release of liquid from the gel matrix to the outer surface) and investigated its interfacial behavior. The oil layer caused by syneresis from the SLUGs at −20 °C was found to show resistance to adhesion of universally adhesive PDA".

From Hönes et al. (2020, *Langmuir* **36**, 4496–4502)

The graphical abstract shows polydopamine with its phenolic hydroxyl groups as stated in the first sentence of the textual abstract. The second sentence of the textual abstract talks about SLUG which is similarly shown in the graphical abstract. The third sentence in the textual

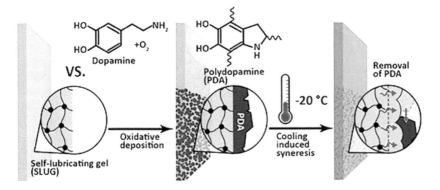

FIGURE 7.3 An example of a graphical abstract having a redundant modal relationship with the textual abstract. The graphical abstract contains a sketch that shows the adsorption of the universal adhesive (polydopamine, PDA) on an oil-infused self-lubricating material. However, when the infused oil is released (–20°C), PDA desorbs from the material. From Hönes et al. (2020, *Langmuir* **36**, 4496–4502).

abstract gives the research question which is answered in the fourth sentence as illustrated in the middle section of the graphical abstract. Finally, the last sentence in the textual abstract is captured in the right portion of the graphical abstract. This clearly shows that the textual and the graphical abstracts have identical contents.

- *Complementary modal relationship* – The graphical and the textual abstracts work together in synergy to communicate a message. An example of a graphical abstract with a complementary modal relationship with the textual abstract is shown in Figure 7.4 (Liu et al. 2020). Liu et al. (2020) describe the design of two isomeric naphthalene appended glucono derivatives *via* substitution at the 1- or 2-naphthyl positions (Nap-**1** and Nap-**2**) as well as their corresponding behaviors in terms of self-assembly and optical properties. The textual abstract says:

"Two isomeric naphthalene appended glucono derivatives substituted at the 1 or 2-naphthyl positions (Nap-**1** and Nap-**2**) were designed and their self-assembly behaviors and optical properties were investigated. Nap-**1** and Nap-**2** were found to self-assemble into nanofibers and nanotwists, respectively. While the molecular chirality of the glucono moiety could not be effectively transferred to the naphthalene moiety in the Nap-**1** system, this was achieved in the Nap-**2** assembly. Thus, the Nap-**2** assembly showed obvious circular dichroism (CD) and circularly polarized luminescence (CPL) signals. From the XRD patterns and IR spectra of the supramolecular assemblies, it was found that Nap-**2** packed in a more orderly fashion than Nap-**1**, leading to a hierarchical assembly forming nanotwist structures. Moreover, a light-harvesting system based on Nap-**2** supramolecular gels and dyes was established, in which an efficient energy transfer was demonstrated from Nap-**2** to an acceptor Eosin Y. It was further found that both chirality and energy transfer enhanced the dissymmetry factor of Eosin Y CPL emission".

From Liu et al. (2020, *Soft Matter* **16**, 4115–4120)

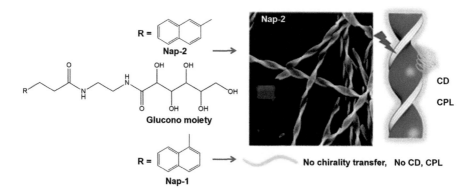

FIGURE 7.4 An example of a graphical abstract having a complementary modal relationship with the textual abstract. The graphical abstract shows the chemical structure of the glucono moiety, positions 1 and 2 on the naphthyl ring as well as a sketch and an image of the corresponding nanofibers and nanotwists they respectively form *via* self-assembly after the glucono moiety attaches at these positions. From Liu et al. (2020, *Soft Matter* **16**, 4115–4120).

The graphical abstract complements the first-three sentences of the textual abstract. Without these sentences, it would have been difficult to understand the graphical abstract, especially trying to work out the 1- and 2-naphthyl positions on the naphthalene ring. The 1- and 2-naphthyl positions (Nap-**1** and Nap-**2**) as well as the glucono moiety mentioned in the first sentence are shown in the left part of the graphical abstract. Nanofibers and nanotwists, formed respectively by Nap-**1** and Nap-**2**, stated in the second sentence, are shown in the middle part of the graphical abstract. Next, the chirality of the nanotwists structures (third sentence) is shown in the right part of the graphical abstract. The other part (sentences four to seven) of the textual abstract then builds on the synergistic foundation laid by sentences one to three and the graphical abstracts to complete the message.

- *Supplementary modal relationship* – The graphical abstract supports the textual abstract (or *vice versa*) through elaboration or exemplification of the content of the other as shown in Figure 7.5. Here, the authors have used the theoretical framework of Rolf Huisgen to account for the elusive existence of the short-lived intermediate $[V_2O]^+$ formed in CO_2 oxidation of $[V_2]^+$ to form $[V_2O_2]^+$. During the oxidation process, $[V_2]^+$ reacts with CO_2 to form the intermediate $[V_2O]^+$ which instantly reacts with CO to form $[V_2O_2]^+$ (Li et al. 2020). This is the idea exemplified in the metaphoric animated cartoon shown in Figure 7.5.
- *Stage-setting modal relationship* – The graphical abstract is used to introduce the textual abstract or *vice versa*, thereby making the reader to look forward to the complete discussion. Figure 7.6, from Volkov et al. (2019), is a typical example of a stage-setting graphical

FIGURE 7.5 An example of a graphical abstract with a supplementary modal relationship with the textual abstract. The graphical abstract is an animated cartoon that explains the short-lived nature of the intermediate $[V_2O]^+$ species formed in the thermal gas-phase oxidation of V_2^+ species by CO_2 to $[V_2O_2]^+$ species using the theoretical framework of Rolf Huisgen. Initially, V_2^+ is oxidized by CO_2 to form the highly reactive $[V_2O]^+$ species which immediately combines with CO in a fast reaction to form the stable $[V_2O_2]^+$ species. From Li et al. (2020, *Angew. Chem. Int. Ed.* **59**, 12308–12314).

FIGURE 7.6 An example of a graphical abstract with stage-setting modal relationship with the textual abstract. This graphical abstract is a photograph of two different plant species in two different pots connected by some sort of cord. Because the graphical abstract is not supported by text, its meaning is somewhat obscured without the textual abstract. In other words, the graphical abstract sets the stage for the textual abstract. From Volkov et al. (2019, *Bioelectrochemistry*, **129**, 70–78).

abstract. It shows two different plants species connected by some sort of chord without any text to explain what could be happening, thereby making the reader to read the textual abstract for the complete message. The textual abstract says:

> "Plants can communicate with other plants using wireless pathways in the plant-wide web. Some examples of these communication pathways are (1) volatile organic compounds' emission and sensing; (2) mycorrhizal networks in the soil; (3) the plants' rhizosphere; (4) naturally grafting of roots of the same species; (5) electrostatic or electromagnetic interactions and (6) acoustic communication. There is an additional pathway for electrical signal transmission between plants - electrical signal transmission between roots through the soil. To avoid the possibility of communication between plants using mechanisms (1) to (6), soils in pots with plants were connected by Ag/AgCl or platinum wires. Electrostimulation of *Aloe vera*, tomato or cabbage plants induces electrotonic potentials transmission in the electro-stimulated plants as well as the plants located in different pots regardless if plants are the same or different types. The amplitude and sign of electrotonic potentials in electro-stimulated and neighboring plants depend on the amplitude, rise and fall of the applied voltage. Experimental results displayed cell-to-cell electrical coupling and the existence of electrical differentiators in plants. Electrostimulation by a sinusoidal wave induces an electrical response with a phase shift. Electrostimulation serves as an important tool for the evaluation of mechanisms of communication in the plant-wide web".
>
> From Volkov et al. (2019, *Bioelectrochemistry*, **129**, 70–78)

iv. **Mediated context:** This classification is based on the way readers encounter graphical abstracts in different journals based on their position and presentation. The role and position of graphical abstracts are not fixed across journals as evidenced by the different ways they are positioned and presented, be it online or in reprint. Therefore, readers may encounter graphical abstracts:

- As a *dominant* presentation of the paper's essence, where the graphical abstract comes before the textual abstract, *e.g.*, as with the *Journal of Organic Chemistry* (Figure 7.7), published by the American Chemical Society.
- In *parallel* with other textual interpretive aids like the textual abstract, keywords or highlights, *e.g.*, in *Science of the Total Environment* (Figure 7.8) published by Elsevier.
- In *subordinate* to textual aids, *e.g.*, embedded within the written abstract as a small figure. This is illustrated in Figure 7.9 from *Chemical Reviews* (published by the American Chemical Society).
- As an *optional* figure when present in some contexts and absent in others or present in online preview and absent in reprint, *e.g.*, in *Chemical Communication* and *Soft Matter*, published by the Royal Society of Chemistry, where the online preview has a graphical abstract with a little descriptive text while the reprinted version does not contain it.

Received June 12, 2008

Heteroatom-Annulated Perylenes: Practical Synthesis, Photophysical Properties, and Solid-State Packing Arrangement

Wei Jiang,[§,‡] Hualei Qian,[§,‡] Yan Li,[§,§] and Zhaohui Wang*,[§]

Beijing National Laboratory for Molecular Sciences, Key Laboratory of Organic Solids, Institute of Chemistry, Chinese Academy of Sciences, Beijing 100190, China, and Graduate School of the Chinese Academy of Sciences, Beijing 100190, China

wangzhaohui@iccas.ac.cn

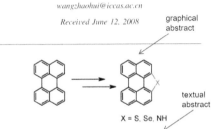

graphical abstract

textual abstract

X = S, Se, NH

A practical strategy for the preparation of a series of heterocyclic annulated perylenes in good yields is presented. UV−vis absorption spectra indicate hypsochromic shift of the absorption maxima relative to the corresponding parent perylene. Single-crystal X-ray diffraction analysis reveals that they all adopt planar conformation, but the solid-state packing arrangements are significantly altered by annulation of various heterocycles.

properties.[3] Incorporating heteroatoms into its skeleton is an intriguing target because the introduction of heteroatoms would induce a variety of intermolecular interactions, such as van der Waals interactions and heteroatom−heteroatom interactions (S⋯S or Se⋯Se interactions), which is essential to achieve highly ordered supramolecular self-assembled structure,[4] and eventually excellent device performance.[5]

S-heterocyclic annulated perylene, namely perylo[1,12-b,c,d]thiophene (PET, **3**), which has first prepared from 3,4:9,10-perylenetetracarboxylic dianhydride by Rogovik,[6] has been synthesized by several groups in harsh conditions such as flash vacuum pyrolysis (FVP).[7] However, its electrical property is rarely studied, probably due to the difficulties in scale-up, long reaction sequences, and poor yields. Recently, we reported its extraordinary solid-state packing arrangement with marked S⋯S short contacts of 3.51 Å between the neighboring columns and the likelihood of double-channel superstructure, which is responsible for effective intermolecular carrier transport.[5c] Herein, we describe our endeavors to develop a new and practical synthetic route toward PET up to gram-scale successfully. Furthermore, Se-heterocyclic annulated perylene is synthesized by incorporating selenium into the perylene skeleton. Accordingly, detailed investigation of photophysical properties and single-crystal analysis of heterocyclic annulated perylenes is presented to fully explore the influence of different heteroatoms on the inherent electronic properties and solid-state packing arrangement.

The key starting material is 1-nitroperylene **2**. Previous synthetic reports[8] of **2** were unsatisfactory in that the yield of the regiospecific mononitration of perylene at position 1 was rather low or the reaction was irreproducible.[9] Successful preparation of **2** from perylene is achieved in a modified mononitration of the perylene process, by which the temperature is reduced and the reaction time is shortened to 25−30 min. There are three main fractions in the crude products. The first

FIGURE 7.7   An example of a graphical abstract with dominant presentation of the paper's essence. The graphical abstract is placed above the textual abstract so that readers encounter it before reading the textual abstract. From Jiang et al. (2008, *J. Org. Chem.* **73**, 7369–7372).

Activity 7.1 *Analysis of Graphical Abstracts into Various Categories*

Analyze the graphical abstracts of published papers in your subject area and place them into the various categories of graphical abstracts discussed in Section 7.2.

7.3   CREATING A GRAPHICAL ABSTRACT

Unlike other sections of a scientific paper with numerous guides, there is very little information on how to create a compelling graphical abstract. At best, journals and publishers give specifications of their individual graphical abstracts either online or in the *Guide for Authors, e.g.*, Elsevier has published specifications for their graphical abstracts online. This notwithstanding, the following steps can be used in combination with publishers' and journals' specifications to create effective and compelling graphical abstracts that serve their sole purpose of summarizing the central message of a paper at a glance.

Science of the Total Environment 575 (2017) 525–535

Contents lists available at ScienceDirect

Science of the Total Environment

journal homepage: www.elsevier.com/locate/scitotenv

Review

Exposure to pesticides and the associated human health effects

CrossMark

Ki-Hyun Kim [a,*], Ehsanul Kabir [b], Shamin Ara Jahan [c]

[a] Department of Civil and Environmental Engineering, Hanyang University, Seoul, 04763, Republic of Korea
[b] Dept. of Farm, Power & Machinery, Bangladesh Agricultural University, Mymensingh, 2202 Bangladesh
[c] BRAC Clinic, Dhaka, 2202 Bangladesh

HIGHLIGHTS

- Pesticides are designed to function with reasonable certainty and minimal risk to human health.
- Pesticide exposure is however turned out to be linked with various diseases including cancer.
- In light of the significance of pesticide pollution, the general aspects of pesticides are assessed.
- The current state of knowledge regarding pesticide use and its detrimental impacts is described.

GRAPHICAL ABSTRACT

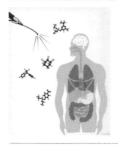

ARTICLE INFO

Article history:
Received 12 July 2016
Received in revised form 21 August 2016
Accepted 1 September 2016
Available online 7 September 2016

Keywords:
Pesticide toxicity
Route of exposure
Environmental effect
Health hazards

ABSTRACT

Pesticides are used widely to control weeds and insect infestation in agricultural fields and various pests and disease carriers (e.g., mosquitoes, ticks, rats, and mice) in houses, offices, malls, and streets. As the modes of action for pesticides are not species-specific, concerns have been raised about environmental risks associated with their exposure through various routes (e.g., residues in food and drinking water). Although such hazards range from short-term (e.g., skin and eye irritation, headaches, dizziness, and nausea) to chronic impacts (e.g., cancer, asthma, and diabetes), their risks are difficult to elucidate due to the involvement of various factors (e.g., period and level of exposure, type of pesticide (regarding toxicity and persistence), and the environmental characteristics of the affected areas). There are no groups in the human population that are completely unexposed to pesticides while most diseases are multi-causal to add considerable complexity to public health assessments. Hence, development of eco-friendly pesticide alternatives (e.g., EcoSMART) and Integrated Pest Management (IPM) techniques is desirable to reduce the impacts of pesticides. This paper was hence organized to present a comprehensive review on pesticides with respect to their types, environmental distribution, routes of exposure, and health impacts.

FIGURE 7.8   An example of a graphical abstract placed in parallel with a textual interpretative aid (highlights). The graphical abstract is a cartoon of a man inhaling sprayed pesticides from the environment in which the pesticides can potentially damage his internal organs. From Kim et al. (2017, *Sci. Total Environ.* **575**, 525–535).

- **Step 1:** Identify the central message of your paper.
- **Step 2:** Select a/figure(s) from the paper or think of a/sketch(es) that best describe(s) the central message and create your graphical abstract around it/them. You can also use figures that are not part of the paper but were obtained during the work.
- **Step 3:** In line with the four-level descriptive taxonomy (Section 7.2), decide the type of graphical abstract you want to create and use the figure(s) or sketches selected to create it.
- **Step 4:** For figures, include sketches and text if necessary and group all the elements into a single figure.

This is an open access article published under an ACS AuthorChoice License, which permits
copying and redistribution of the article or any adaptations for non-commercial purposes.

Review

Cite This: *Chem. Rev.* 2019, 119, 3510−3673

pubs.acs.org/CR

Pharmaceuticals of Emerging Concern in Aquatic Systems: Chemistry, Occurrence, Effects, and Removal Methods

Manvendra Patel,[†] Rahul Kumar,[‡] Kamal Kishor,[‡] Todd Mlsna,[‡] Charles U. Pittman, Jr.,[‡] and Dinesh Mohan*[,‡]

[†]School of Environmental Sciences, Jawaharlal Nehru University, New Delhi 110067, India

[‡]Department of Chemistry, Mississippi State University, Mississippi State, Mississippi 39762, United States

ABSTRACT: In the last few decades, pharmaceuticals, credited with saving millions of lives, have emerged as a new class of environmental contaminant. These compounds can have both chronic and acute harmful effects on natural flora and fauna. The presence of pharmaceutical contaminants in ground waters, surface waters (lakes, rivers, and streams), sea water, wastewater treatment plants (influents and effluents), soils, and sludges has been well doccumented. A range of methods including oxidation, photolysis, UV-degradation, nanofiltration, reverse osmosis, and adsorption has been used for their remediation from aqueous systems. Many methods have been commercially limited by toxic sludge generation, incomplete removal, high capital and operating costs, and the need for skilled operating and maintenance personnel. Adsorption technologies are a low-cost alternative, easily used in developing countries where there is a dearth of advanced technologies, skilled personnel, and available capital, and adsorption appears to be the most broadly feasible pharmaceutical removal method. Adsorption remediation methods are easily integrated with wastewater treatment plants (WWTPs). Herein, we have reviewed the literature (1990−2018) illustrating the rising environmental pharmaceutical contamination concerns as well as remediation efforts emphasizing adsorption.

FIGURE 7.9   An example of a graphical abstract embedded in a textual abstract. The graphical abstract shows that toxic substances from animal, human, industrial and hospital effluents enter the environment with harmful effects. The presence of these substances in the environment can be detected by sampling and analyzing either water, soil or air while their removal from the environment is by an appropriate treatment method. From Patel et al. (2019, *Chem. Rev.* **119**, 3510–3673).

- **Step 5:** Ask yourself if the graphical abstract is eye-catching, summarizes the central message at a glance and can effectively draw readers' attention to the paper? If your answer is "**yes**"; you have succeeded in creating an effective graphical abstract, but if "**no**" edit it carefully until your answer is a resounding "**yes**".
- **Step 6:** Save the graphical abstract in the format and size acceptable by your target journal: *e.g.*, jpeg, png or tiff and 531 × 1328 pixels, *etc.*

As a guide:

- Do not flood your graphical abstract with too many figures and text: use thcm sparingly and leave-out plenty white space.
- Use a good choice of color blend for sketches and text.
- Be sure that the line thickness of sketches and text size are such that the figure will remain legible after compression into the journal space.

- Learn from others: analyze graphical abstracts of papers in your area, paying particular attention to their clarity and effectiveness in conveying the central message at a glance as well as attracting readers' attention and mimic this in your graphical abstracts.

7.4   SOFTWARE FOR CREATING GRAPHICAL ABSTRACTS

The following software can be used to create graphical abstracts:

- Microsoft Office PowerPoint: *Pros* – it is simple to use, has in-built shapes to select from and supports vector graphics, and images can be saved in different formats (*e.g.*, jpeg, png, tiff, *etc.*). *Cons* – it is a paid software and has limited drawing tools.
- Microsoft Paint: *Pros* – it is free (in Windows), is simple to use and supports important image formats. *Cons* – it has limited drawing tools.
- Adobe Photoshop and Illustrator: *Pros* – it is a professional, state-of-the-art software. *Cons* – it is a paid software and requires time and effort to learn, and it does not work on Linux computers.
- CorelDraw (Windows) and Affinity Design (Mac): *Pros* – they are professional and state-of-the-art software. *Cons* – they are paid software, require time and effort to learn and work only on the operating systems they have been designed for.
- InkScape and Gimp: *Pros* – they are free, open-source and semi-professional software. *Cons* – they require time and effort to learn, and they also lack advanced features.
- ChemDraw: *Pros* – it is a professional and state-of-the-art software. *Cons* – it is a paid software whose use is limited to chemical and biochemical structures.

This list cannot be exhausted as more drawing software get developed yearly; therefore, look out for more and choose the one that is most convenient for you.

Activity 7.2 *Creating a Graphical Abstract*

From the foregoing, create a graphical abstract for any completed simple experiment of your choice. This could be a routine or an occasional experiment in your laboratory.

FURTHER READING

Hullman, J., and B. Bach. 2018. "Picturing science: Design patterns in graphical abstracts." International Conference on Theory and Application of Diagrams.
- Discusses the design principle of graphical abstracts in both science and social science as well as the role these abstracts play in disseminating research findings.

Ibrahim, A.M. 2018. Use of a visual abstract to disseminate scientific research.
med.brown.edu/pedisurg/IllustrationClass/VisualAbstract_Primer_v4_1.pdf
- Gives practical suggestions for designing a graphical abstract and discusses the role graphical abstracts perform in disseminating research findings.

REFERENCES

Choi, J., W. Kim, and H.-Y. Kim. 2020. Crack density in bloodstains. *Soft Matter* 16 (24):5571–5576.

Hönes, R., Y. Lee, C. Urata, H. Lee, and A. Hozumi. 2020. Antiadhesive properties of oil-infused gels against the universal adhesiveness of polydopamine. *Langmuir* 36 (16):4496–4502.

Lane, S., A. Karatsolis, and L. Bui. 2015. Graphical abstracts: a taxonomy and critique of an emerging genre. *Proceedings of the 33rd Annual International Conference on the Design of Communication.*

Li, J., C. Geng, T. Weiske, and H. Schwarz. 2020. Counter-intuitive gas-phase reactivities of [V2]+ and [V2O]+ towards CO_2 reduction: Insight from electronic structure calculations. *Angew. Chem. Int. Ed.* 59 (30):12308–12314.

Liu, Z., Y. Jiang, J. Jiang, D. Zhai, D. Wang, and M. Liu. 2020. Self-assembly of isomeric naphthalene appended glucono derivatives: nanofibers and nanotwists with circularly polarized luminescence emission. *Soft Matter* 16 (17):4115–4120.

Ma, Y., A. Sumboja, W. Zang, S. Yin, S. Wang, S.J. Pennycook, Z. Kou, Z. Liu, X. Li, and J. Wang. 2019. Flexible and wearable all-solid-state Al–air battery based on iron carbide encapsulated in electrospun porous carbon nanofibers. *ACS Appl. Mater. Interfaces* 11 (2):1988–1995.

Pferschy-Wenzig, E.-M., U. Pferschy, D. Wang, A. Mocan, and A.G. Atanasov. 2016. Does a graphical abstract bring more visibility to your paper? *Molecules* 21:1247.

Pitto-Barry, A., and N.P.E. Barry. 2019. Effect of temperature on the nucleation and growth of precious metal nanocrystals. *Angew. Chem. Int. Ed.* 58 (51):18482–18486.

Qi, W., J. Li, and P.B. Weisensee. 2019. Evaporation of sessile water droplets on horizontal and vertical biphobic patterned surfaces. *Langmuir* 35 (52):17185–17192.

Tyowua, A.T., F. Ahor, S.G. Yiase, and B.P. Binks. 2020. Liquid marbles as microreactors for qualitative and quantitative inorganic analyses. *SN Appl. Sci.* 2 (3):345.

Volkov, A.G., S. Toole, and M. WaMaina. 2019. Electrical signal transmission in the plant-wide web. *Bioelectrochemistry* 129:70–78.

Zhang, C., H. Deng, S.M. Kenderes, J.-W. Su, A.G. Whittington, and J. Lin. 2019. Chemically interconnected thermotropic polymers for transparency-tunable and impact-resistant windows. *ACS Appl. Mater. Interfaces* 11 (5):5393–5400.

8 Highlights

8.1 FUNCTIONS OF HIGHLIGHTS

Like the graphical abstract, highlights give an overview and a concise summary of the core findings of the paper at a glance (Cagliero and La Quatra 2020). Highlights serve as the *elevator pitch* of the paper, with the primary aim of arousing readers' interest. Highlights also increase visibility and discoverability of the paper in online searches – both within and outside the immediate research community – thereby increasing the chances of collaborations. Highlights were introduced in scholarly publishing by Elsevier, around 2010, in the form of bullet-point sentences (≤ 85 characters, including spaces). With the aim of making research papers readily accessible on smartphones, Elsevier launched the "Research Highlights App" in January 2014. Nowadays, publishers like the Royal Society of Chemistry and Wiley use a brief text, published online with the graphical abstract, rather than bullet-point sentences, as highlights.

8.2 WRITING HIGHLIGHTS

Depending on the journal, highlights may be submitted (as bullet points or a brief text) along with the manuscript during the peer-review process or may be required after the paper has been accepted. Whether in bullet points or a brief text:

- Be sure that your highlights contain the essence of the paper – *i.e.*, results or conclusions. Specifically, include the (i) nature of the research, *i.e.*, the reason *why* the work was done; (ii) key findings; and (iii) contribution of the work to the field. However, do not include the background and the methodology.
- Write highlights in present tense using active sentences. The lead-in or first sentence should state the *nature* of the work. Thereafter, state the *key finding(s)* and then finally, state the *contribution* or main *conclusion*.
- Use keywords, but write for a general audience, *i.e.*, "everyone in the world".
- Highlights should be able to standalone; therefore, avoid subject-specific terms, acronyms and abbreviations.

Example 8.1 *Highlights of a Scientific Research Paper*

- Plants can communicate with other plants using pathways in the plant-wide web.

DOI: 10.1201/9781003186748-8

- There is a pathway for the electrical signaling between roots through the soil.
- Electrostimulation induces electrotonic potentials in the neighboring plants.
- Signal transmission can be between the same or different types of plants.
- There are electrical differentiators and cell-to-cell electrical coupling in plants.

From Volkov et al. (2019, *Bioelectrochemistry*, **129**, 70–78)

Example 8.2 *Highlights of a Scientific Review Paper*

- Pesticides are designed to function with reasonable certainty and minimal risk to human health.
- Pesticide exposure is however turned out to be linked with various diseases including cancer.
- In light of the significance of pesticide pollution, the general aspects of pesticides are assessed.
- The current state of knowledge regarding pesticide uses and its detrimental impacts are described.

From Kim et al. (2017, *Sci. Total Environ.* **575**, 525–535)

Activity 8.1 *Recognizing the Use of Present Tense in the Highlight Section of Scientific Papers*

Analyze the highlight section of at least five published papers in your subject area. Are the sentences in present or past tense? What is the minimum and maximum number of words contained in the highlight section?

Activity 8.2 *Writing a Highlight Section*

From the foregoing, write the highlight for any completed simple experiment of your choice. This could be a routine or an occasional experiment in your laboratory.

FURTHER READING

https://www.elsevier.com/authors/tools-and-resources/highlights
- Gives the requirements for writing the highlight section of a scientific paper for journals published by Elsevier.

https://www.biosciencewriters.com/Tips-for-Writing-a-Highlights-Section-of-a-Scientific-Manuscript.aspx
- Contains tips for writing the highlight section of a scientific paper.

https://www.letpub.com/What-are-Manuscript-Highlights
- Contains useful suggestions for writing the highlight section of a research paper.

REFERENCES

Cagliero, L., and M. La Quatra. 2020. Extracting highlights of scientific articles: A supervised summarization approach. *Expert Syst. Appl.* 160: 113659.

Kim, K.-H., E. Kabir, and S.A. Jahan. 2017. Exposure to pesticides and the associated human health effects. *Sci. Total Environ.* 575:525–535.

Volkov, A.G., S. Toole, and M. WaMaina. 2019. Electrical signal transmission in the plant-wide web. *Bioelectrochemistry* 129:70–78.

9 Introductory Section of a Research Paper

9.1 FUNCTIONS OF THE INTRODUCTORY SECTION OF A RESEARCH PAPER

The functions of the introductory section of a research paper are in two folds: (i) to provide a convincing justification for the work and (ii) to give a clear aim or hypothesis of the work. According to Swales and Feak (2004), the introductory section of a research paper contains three *moves* (*i.e.*, three broad subsections) which are organized sequentially or randomly. These subsections are not clearly marked out in the introductory section of a research paper, but they can be identified by close examination.

- **Move 1** – The author establishes the research area by:
  Optional
  a. Showing that the general research area is important, central, interesting, problematic or relevant in some way.
  Obligatory
  b. Introducing and reviewing previously published materials on the subject area.
- **Move 2** – (*Obligatory*) The author establishes a niche or a research gap by pointing out a weakness or weaknesses in previously published materials reviewed in move 1. This could be a problem, a question or an extension of previous knowledge.
- **Move 3** – The author occupies the niche or fills the research gap by:
  Obligatory
  a. Stating how (i) the problem can be fixed, (ii) the question can be answered or (iii) previous knowledge can be extended.
  b. Outlining the aim of the research or stating the nature of the research.
  Optional
  c. Listing the research questions or hypotheses to be tested.
  d. Announcing the key findings.
  e. Stating the importance of the research.
  f. Or stating the structure of the research paper.

An elaborate explanation of these moves can be found in Lewin et al. (2001). These moves ensure that the introductory section of a research paper necessarily contains (i) the background information about the work, including what is known and what is unknown; (ii) the importance of the work; and (iii) the aim of the work, the problem

DOI: 10.1201/9781003186748-9

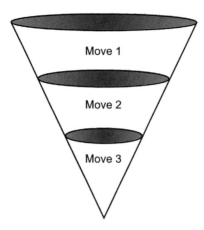

FIGURE 9.1   Schematic illustration of the inverted "cone-shape" nature of a move-based introductory section of a research paper. Move 1 is placed in the relatively broad part of the cone because it contains a broad overview of the topic. The topic is narrowed down in move 2 as indicated by the relatively small portion it occupies in the middle of the cone. Move 3 contains very specific statements about the topic, thus occupying the smallest portion of the cone.

statement and the research question or hypothesis. These moves (1–3) give the introductory section a "cone-shape" because the contents are from a "general background knowledge to a specific problem or aim" (Cals and Kotz 2013b). It is the opposite of the "inverted cone-shape" of the discussion section (Chapter 12) which goes from "specific results to a general conclusion" (Cals and Kotz 2013a).

9.2   WRITING A MOVE-BASED INTRODUCTORY SECTION OF A RESEARCH PAPER

Examine your key finding(s) critically, get the central message clearly and then structure and write the introduction around this message (Cals and Kotz 2013b).

a. **Move 1 – Establishing the research area**
   Begin broadly by either (i) stating the general knowledge or acceptable facts about the area, (ii) stating the importance of the work or research area, (iii) giving the background information about the work or research area, (iv) giving a general description of the area, (v) defining keywords associated with the area or (vi) a mixture of (i)–(v). Thereafter, narrow down to the specific theme of the paper and review specific relevant literature, stating both what is known and what is unknown (or problems). Problems with previous research must be stated professionally rather than arrogantly, provocatively and/or confrontationally.

b. **Move 2 – Establishing a niche or research gap**
   New research is done because there is some gap in knowledge. The gap may either be (i) that previous research can be extended, (ii) a solution to a problematic

method, (iii) a new mathematical model that describes a phenomenon better, (iv) a new theory that needs to be tested or (v) an invention of a new material or product. Whatever the knowledge gap, establish your niche or research gap in line with what you have claimed is unknown (or a problem) about the area. Articulate your niche succinctly in a single sentence. Next, state clearly why the niche requires attention; in other words, why the work is needed.

c. **Move 3 – Occupying the niche or filling the research gap**
State clearly how you plan to investigate the niche or close the research gap identified by stating the aim and plan of the work. However, do not give out the result here – save it for the "results" or the "results and discussion section". Some journals may require a summary of the findings – consult the *Guide for Authors* or recent volumes of the journal for possible clues. These moves are illustrated in Example 9.1.

Example 9.1 *Analysis of the Introductory Section of a Research Paper*

Analyze and identify the various moves in the introductory section of Mondragón-Cortez et al. (2013) entitled "Physicochemical characterization of honey from the West region of México".

INTRODUCTION

"Honey is a natural complex food product produced by bees from the nectar of plants. Bee honey has significant nutritional and medicinal benefits. It is a rich source of readily available sugars, organic acids, various amino acids and in addition source of many biologically active compounds.

Each honey possesses a unique combination of components and properties, due to the variety of flora located in its geographical region of production, climatic conditions of the producing area and processing and storage methods.

The physical properties and chemical composition of honey from different countries have been published by many researchers. Honeys also are analyzed for the pollen content and to control their origin. This kind of analysis has become more popular in recent years since the characterization of honey, one of the most highly valued apicultural products, is an important aspect with respect to bee-keeping development.

Mexican beekeeping has a high social and economic value. Currently, Mexico is the fifth largest producer of honey with about 57,000 tons and the third largest exporter in the World. México exports honey principally to Germany, England and the United States, but the problem is that generally beekeepers are selling their honey without a characterization.

Some studies have reported the physicochemical characteristics of honey from the Southwest of México; however, there is no reported information from other regions of the country. Therefore, the generation of information on honey from regions of México where its characteristics are unknown, could be useful for the integration of a national map of honey quality and its comparison with the international standards and parameters of honey from other countries. The objective of this research is physicochemical characterization and pollen quantification of honey from the West region of México".

From Mondragón-Cortez et al. (2013, *CyTA - J. Food* **11**, 7–13)

ANALYSIS

MOVE 1: GENERAL KNOWLEDGE

"Honey is a natural complex food product produced by bees from the nectar of plants. Bee honey has significant nutritional and medicinal benefits. It is a rich source of readily available sugars, organic acids, various amino acids and in addition source of many biologically active compounds".

MOVE 1: ACCEPTABLE FACT

"Each honey possesses a unique combination of components and properties, due to the variety of flora located in its geographical region of production, climatic conditions of the producing area and processing and storage methods".

MOVE 1: LITERATURE REVIEW (ON PROPERTIES OF HONEY)

"The physical properties and chemical composition of honey from different countries have been published by many researchers. Honeys also are analyzed for the pollen content and to control their origin. This kind of analysis has become more popular in recent years since the characterization of honey, one of the most highly valued apicultural products, is an important aspect with respect to beekeeping development".

MOVE 1: STATING THE PROBLEM

"Mexican beekeeping has a high social and economic value. Currently, Mexico is the fifth largest producer of honey with about 57,000 tons and the third largest exporter in the World. México exports honey principally to Germany, England and the United States, but the problem is that generally beekeepers are selling their honey without a characterization".

MOVE 2: ESTABLISHING THE NICHE OR RESEARCH GAP

"Some studies have reported the physicochemical characteristics of honey from the Southwest of México; however, there is no reported information from other regions of the country. Therefore, the generation of information on honey from regions of México where its characteristics are unknown, could be useful for the integration of a national map of honey quality and its comparison with the international standards and parameters of honey from other countries".

MOVE 3: OCCUPYING THE NICHE OR FILLING THE RESEARCH GAP

"The objective of this research is physicochemical characterization and pollen quantification of honey from the West region of México".

Activity 9.1 *Analyzing the Introductory Section of a Research Paper*

Analyze and identify the various moves in the introductory section of Arnell and Reynard (1996) entitled "The effects of climate change due to global warming on river flows in Great Britain".

INTRODUCTION

"Increasing concentrations of greenhouse gases in the atmosphere are predicted to result in an increase in global mean temperature of the order of 0.3 °C per decade, in the absence of preventive strategies. Such a global warming could have a significant impact on local and regional climatic regimes, which would in turn impact upon hydrological and water resources systems.

Over the last decade, there have been many investigations into the sensitivity of hydrological regimes to climatic changes associated with global warming, in a wide range of environments and using many different models and scenarios.

This paper presents some results of an investigation into potential changes in river flow regimes in Great Britain. The broad objective was to simulate possible changes in ever flows in a wide range of catchments, examining changes in annual and monthly runoff and low flow extremes. The study used a conceptual daily rainfall-runoff model, a baseline period of 1951 to 1980 and equilibrium and transient climate change scenarios derived from the output from general circulation models.

From Arnell and Reynard (1996, *J. Hydrol.* **183**, 397–424)

9.3   USE OF TENSES AND REFERENCES IN THE INTRODUCTORY SECTION OF A RESEARCH PAPER

Tenses for Move 1

- Use the *present tense* to state established knowledge, *e.g.*, "honey is a natural complex food product produced by bees from the nectar of plants" (Example 9.1, established knowledge).
- Use the *present perfect tense* to refer to previous research (literature review), especially if the research is recent, *e.g.*, "the physical properties and chemical composition of honey from different countries have been published by many researchers" (Example 9.1, literature review). Also, use *present perfect tense* to lay emphasis on a finding rather than the author.
- Use the *past tense* to lay emphasis on an author in relation to a finding.
- Use a mixture of both passive and active sentences. Use passive sentences to lay emphasis on the findings of a researcher and active sentences to lay emphasis on the researcher who reported the findings. This is achieved using the appropriate verb tenses. For instance, "the preparation of non-wettable carbon nanotubes has also been reported" (Lau et al. 2003) is a *passive sentence* formed by using the *present perfect tense*. "Lau et al. (2003) have reported the preparation of non-wettable carbon nanotubes" is an *active sentence* formed by using the *simple past tense*. The first sentence emphasizes the preparation of non-wettable carbon nanotubes, while the second one emphasizes the researchers who prepared the carbon nanotubes.

Tense for Move 2

- Use the present perfect tense + signal words like "however", "but" or "nonetheless" to state a research gap or stress a niche. For example, "some studies have reported the physicochemical characteristics of honey from the Southwest of México; however, there is no reported information from other regions of the country" (Example 9.1, establishing a niche).

Tense for Move 3

- Use the *simple present tense* to state the aim of your research, *e.g.*, "the objective of this research is physicochemical characterization and pollen quantification of honey from the West region of México" (Example 9.1, occupying a niche).

Referencing

- Reference all important statements, with their original sources, especially when reviewing papers related to your work. Different referencing methods – information prominent method, author prominent method or weak author prominent method (Cargill and O'Connor 2013) – can be used. Consult the *Guide for Authors* of your target journal for the appropriate referencing method.

Information prominent method

- The reference is placed at the end of the sentence either as author and year or number (as superscript or in square bracket), *e.g.*, "the particle shell encasing a liquid drop is permeable to gas and can be used for gas sensing in environmental science (Tian et al. 2010)".

Author prominent method

- The author is placed prominently at the beginning of the sentence, *e.g.*, "Tian et al. (2010) showed that the particle shell encasing a liquid drop is permeable to gas and can be used for gas sensing in environmental science".

Weak author prominent method

- This method is similar to the information prominent method; however, the statement is attributed to two or more authors which are listed at the end of the statement. Thereafter, a statement highlighting the contribution of one of the authors is given. For example, "the particle shell encasing a liquid drop is permeable to gas and can be used for gas sensing in environmental science (Tian et al. 2010, Bormashenko 2011, Tyowua et al. 2020). For instance, Tian et al. (2010) showed that ..."

Activity 9.2 *Analyzing and Identifying Various Moves in the Introductory Section of Scientific Papers*

Analyze and identify the various moves in the introductory section of a least five published research papers in your area of specialty. Also, identify the various tenses used in each move as well as the referencing styles. You will notice that they are written in line with Sections 9.1–9.3.

FURTHER READING

Mack, C.A. 2012. How to write a good scientific paper: title, abstract, and keywords. *J. Micro-Nanolith. MEMS.* 11 (2):020101.
  • Contains useful suggestions for writing the introductory section and various parts of a scientific research paper. Scientific publishing ethics, peer-review, plagiarism and other aspects of scientific publishing are also discussed.
Weissberg, R., and S. Buker. 1990. Writing up research: Prentice Hall Englewood Cliffs, NJ.
  • An excellent guide for writing up all the sections of a scientific research paper. Useful suggestions for writing a compelling introductory section of a research paper are also given.

REFERENCES

Arnell, N.W., and N.S. Reynard. 1996. The effects of climate change due to global warming on river flows in Great Britain. *J. Hydrol.* 183 (3):397–424.

Bormashenko, E. 2011. Liquid marbles: Properties and applications. *Curr. Opin. Colloid Interface Sci.* 16 (4):266–271.

Cals, J.W., and D. Kotz. 2013a. Effective writing and publishing scientific papers, part VI: discussion. *J. Clin. Epidemiol.* 66 (10):1064.

Cals, J.W.L., and D. Kotz. 2013b. Effective writing and publishing scientific papers, part III: introduction. *J. Clin. Epidemiol.* 66 (7):702.

Cargill, M., and P. O'Connor. 2013. *Writing scientific research articles: Strategy and steps*: John Wiley & Sons, Essex.

Lewin, B., J. Fine, and L. Young. 2001. *Expository discourse: A genre-based approach to social science research texts*: Continuum, London.

Mondragón-Cortez, P., J.A. Ulloa, P. Rosas-Ulloa, R. Rodríguez-Rodríguez, and J.A. Resendiz Vázquez. 2013. Physicochemical characterization of honey from the West region of México. *CyTA - J. Food* 11 (1):7–13.

Swales, J.M., and C.B. Feak. 2004. *Academic writing for graduate students: Essential tasks and skills*. Vol. 1: University of Michigan Press Ann Arbor, MI.

Tian, J., T. Arbatan, X. Li, and W. Shen. 2010. Liquid marble for gas sensing. *Chem. Commun.* 46 (26):4734–4736.

Tyowua, A.T., F. Ahor, S.G. Yiase, and B.P. Binks. 2020. Liquid marbles as microreactors for qualitative and quantitative inorganic analyses. *SN Appl. Sci.* 2 (3):345.

10 Materials and Methods Section of a Research Paper

10.1 FUNCTIONS OF THE MATERIALS AND METHODS SECTION

Depending on the journal, the "Materials and Methods" section of a research paper is also referred to as "Methods", "Methodology", "Experimental" or "Experimental Procedures". The section is subdivided into the "Materials" and the "Methods" subsections. The "Materials" subsection contains a list of all chemical reagents (with their purity and name of manufacture) used in the work as shown in Example 10.1 (from Li et al. 2020). The materials: chicken eggs, sodium dihydrogen phosphate, di-sodium hydrogen phosphate, sodium azide and water used in the work are given with their various sources. The "Methods" subsection contains a detailed step-by-step report of all experiments performed in the work including their duration, concentrations, experimental conditions, names, model and the manufacturer of all the equipment used. A typical method subsection is given in Example 10.1 (Li et al. 2020). The time taken for the experiment, the concentration of reagents, the experimental conditions, the names, the model and the manufacture of all the equipment used are given. The information provided in the "Materials and Methods" section helps referees and readers determine whether or not the results reported are credible. Additionally, the information is useful for independent replication and evaluation of the results.

Example 10.1 *Materials and Methods Section of a Paper*

MATERIALS

"Chicken eggs were purchased from a local supermarket (Tesco Ltd., UK). Sodium dihydrogen phosphate, di-sodium hydrogen phosphate, and sodium azide were purchased from Sigma-Aldrich (Dorset, UK). Water purified by treatment with a Milli-Q apparatus (Millipore, Bedford, UK), with a resistivity not less than 18.2 MΩ cm at 25°C was used for the preparation of phosphate buffer. The latter was used as the solvent throughout the experiments with addition of 0.02 wt. % sodium azide as a bactericide".

DOI: 10.1201/9781003186748-10

METHODS

i. *Preparation of egg white protein dispersion*
"Egg white was extracted from the yolk of freshly purchased eggs manually and then homogenized under magnetic stirring (500 rpm speed) for 2 h, as reported previously (Li et al., 2019). No further purification of the egg white protein dispersion was performed".

ii. *Preparation of microgels*
"Egg white protein microgels were prepared *via* a top-down approach of preparing heat-set protein hydrogel followed by controlled shearing using a previous technique with some modifications (Sarkar et al. 2017). Briefly, a 6.25 wt. % egg white protein dispersion, obtained by diluting the egg white protein in 20 mM phosphate buffer at pH 7.0 was thermally-cross-linked by heating (quiescent) at 90 °C for 30 min in a water bath (under quiescent conditions). The gel was then broken up into coarse pieces and passed (twice) through the Leeds Jet homogenizer (University of Leeds, UK) at 300 bar".

From Li et al. (2020, *Food Hydrocoll.*, **98**, 105292).

10.2   WRITING THE MATERIALS AND METHODS SECTION

- Use *simple past tense* in *passive voice* – *i.e., past simple passive (e.g., "…* was determined") – to report what you have done (Example 10.1). The use of *present tense* in *passive voice* is also possible – *i.e., present simple passive (e.g., "…* is determined"). Some journals allow the use of *first-person plural (e.g.,* "we determine …" as in Example 10.2). Check the *Guide for Authors* or recent volumes of the journal you are targeting for clues on the appropriate tenses for the materials and methods section. Earlier studies by Riley (1991), Ding (2002) and Ping Alvin (2014) showed that the materials and methods section of scientific research papers are largely written in past simple passive. However, a later study by Ping Alvin (2020) reported a decline in the use of past simple passive and an increase in the use of active sentences, with the first-person plural "we".
- Use the *simple present tense* to refer to figures and tables that help explain what you have done, *e.g.,* "Figure 1 shows the various stages for extracting pollen in honey" (Lutier and Vaissière 1993).
- State the importance of each experiment to your work, including any inter-relationships between them, and explain why you preferred one method over another where multiple methods are known.
- Reference methods described previously, but leave out their details as illustrated in Example 10.1. However, cite the source and give the details if (i) the source is not readily accessible, (ii) the source is not written in English or (iii) you have modified the method.
- Give as much details as possible, if the method is novel or invented by you. As shown in Example 10.1, details that must be included in the method are quantities, concentrations, experimental conditions (*e.g.,* temperature and pressure),

duration and frequency of measurement, equipment used (including model and name of manufacturer), statistical analyses (including tools, name of software used as well as confidence levels) and sampling location (for field work). Sampling locations are best described with maps (see Section 10.3).

- For clarity, support the text of lengthy and complex methods with flowcharts, figures or tables. Also, use a table instead of text to give the names, purity and source of chemical reagents where more than five chemical reagents are used (see examples in Binks et al. 2014 and Tyowua et al. 2020).
- Place your work in positive light by describing your method positively, stating how careful and accurate you were.
- Report steps, sequentially as they were performed in a given experiment. For example, "sodium chloride (10 g) was weighed into a 250 mL-conical flask and distilled water (150 mL) was added to dissolve it" (*acceptable*). "Distilled water (10 mL) was added to a 250 mL-conical flask to which sodium chloride (10 g) was weighed into so as to dissolve it" (*unacceptable*).
- Like in Examples 10.1 and 10.2, the verb "was" (singular) or "were" (plural) is used in every passive sentence of a materials and methods section. Therefore, pay particular attention to these verbs and use them appropriately – use "was" for a singular event and "were" for plural events.
- Include an ethics statement if humans, animals, stem cells or biohazardous materials were used. Some journals require an "ethics committee approval certificate" and "permission to publish" before considering a submitted manuscript for publication if humans, animals, stem cells or biohazardous materials are used.

Example 10.2 *Methods Section of a Paper with Passive (P) and Active (A) Sentences*

"We used two honey standards (**A**). The first one was extracted in August 1991 from a single honey-bee colony in Neuvic (650 m altitude, 45°21′40″N and 2°20′16″E) and is referred to as 'mountain' honey in the following (**P**). The second standard, referred to as 'test' honey, was made from pollen-free honey filtered and bottled by Burleson and Sons (Waxahachie, TX, USA) to which pollen from *Castanea*, *Helianthus,* and *Lilium* was added to give a mixture with an overall density of 945, 000 pollen grains per 10 g made of 55.4% *Castanea*, 40.5% *Helianthus* and 4.1% *Lilium* pollen grain-wise (**P**). This test honey with its precisely known pollen concentration spectrum of species provided a powerful tool to evaluate pollen analysis method experimentally (**P**)".

From Lutier and Vaissière (1993, *Rev. Palaeobot. Palyno.*, **78**, 129–144).

Activity 10.1 *Changing Passive Sentences to Active Sentences*

Using Example 10.1, change the passive sentences to active sentences with first-person pronouns and active sentences with pronouns to passive sentences without pronouns. You will find out that doing this does not change the key contents.

10.3   USING GEOGRAPHICAL MAPS IN MATERIALS AND METHODS SECTION

A map is a diagrammatical representation of an area, *e.g.*, the geographical location of a field work (Figure 10.1). Maps are expected to "standalone" like figures and tables. As a guide, your map should be labeled as a "figure" and like in Figure 10.1, it should contain the following:

- Latitudes and longitudes.
- Scale bar.
- Sample collection points.
- A legend that explains all symbols used.
- And a caption that contains the name of the map as well as its source. One common error in map caption is leaving out the name of the map, *e.g.*, "geology map of the study area and water sampling points ..." is inappropriate because the name of the study area is not given. However, "geology map

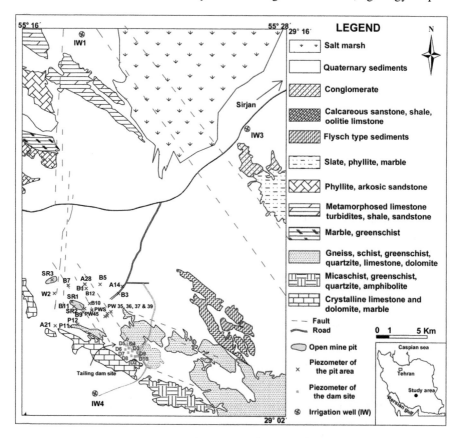

FIGURE 10.1   Geology map of Golgohar (Southern Iran) and water sampling points – modified from geology map of 1:100,000 of the Golgohar sheet. From Jahanshahi and Zare (2015, *Environ. Earth Sci.*, **74**, 505–520).

of Golgohar (Southern Iran) and water sampling points …" is appropriate because the name of the study is stated. This is illustrated in Figure 10.1 (from Jahanshahi and Zare 2015). The map will not standalone once the name of the study area is left out in the caption.

FURTHER READING

A handbook for authors: Preparing your manuscript for Oxford University Press. https://global.oup.com/academic/pdf/authors/authors-handbook.pdf
- Discusses the use of geographical maps in scientific writings and contains useful links to libraries of geographical maps.

Glasman-Deal, H. 2020. *Science Research Writing: for native and non-native speakers of English*: Imperial College Press, London.
- Discusses the language style and structure for writing a scientific research paper and contains an excellent guide for writing the materials and methods section.

Mack, C.A. 2012. How to write a good scientific paper: title, abstract, and keywords. *J. Micro-Nanolith. MEMS.* 11 (2):020101.
- Contains useful suggestions for writing the materials and methods section and various parts of a scientific research paper.

Weissberg, R., and S. Buker. 1990. Writing up research: Prentice Hall Englewood Cliffs, NJ.
- An excellent guide for writing up all the sections of a scientific research paper with excellent suggestions for writing the materials and methods section.

REFERENCES

Binks, B.P., T. Sekine, and A.T. Tyowua. 2014. Dry oil powders and oil foams stabilised by fluorinated clay platelet particles. *Soft Matter* 10 (4):578–589.

Ding, D.D. 2002. The Passive Voice and Social Values in Science. *J. Tech. Writ. Commun.* 32 (2):137–154.

Jahanshahi, R., and M. Zare. 2015. Assessment of heavy metals pollution in groundwater of Golgohar iron ore mine area, Iran. *Environ. Earth Sci.* 74 (1):505–520.

Li, X., J. Li, C. Chang, C. Wang, M. Zhang, Y. Su, and Y. Yang. 2019. Foaming characterization of fresh egg white proteins as a function of different proportions of egg yolk fractions. *Food Hydrocoll.* 90:118–125.

Li, X., B.S. Murray, Y. Yang, and A. Sarkar. 2020. Egg white protein microgels as aqueous Pickering foam stabilizers: Bubble stability and interfacial properties. *Food Hydrocoll.* 98: 105292.

Lutier, P.M., and B.E. Vaissière. 1993. An improved method for pollen analysis of honey. *Rev. Palaeobot. Palyno.* 78 (1):129–144.

Ping Alvin, L. 2014. The passive voice in scientific writing. The current norm in science journals. *J. Sci. Commun.* 13 (1):A03.

Ping Alvin, L. 2020. The passive voice in scientific writing through the ages: A diachronic study. *Text & Talk* 1 (ahead-of-print).

Riley, K. 1991. Passive voice and rhetorical role in scientific writing. *J. Tech. Writ. Commun.* 21 (3):239–257.

Sarkar, A., F. Kanti, A. Gulotta, B.S. Murray, and S. Zhang. 2017. Aqueous lubrication, structure and rheological properties of whey protein microgel particles. *Langmuir* 33 (51):14699–14708.

Tyowua, A.T., F. Ahor, S.G. Yiase, and B.P. Binks. 2020. Liquid marbles as microreactors for qualitative and quantitative inorganic analyses. *SN Appl. Sci.* 2 (3):345.

11 Results Section of Research Papers

11.1 FUNCTIONS OF THE RESULTS SECTION OF A RESEARCH PAPER

The functions of the results section are (i) to show the results of all the experiments performed in the experimental section and (ii) to state any trends emerging from the results. Some journals combine the results section with the discussion section. In this case, the results and the trends are discussed vividly including their ramifications and limitations. The results are presented either in the form of photographs, sketches, flowcharts, tables or graphs.

11.2 RESULTS: PHOTOGRAPHS, SKETCHES, FLOWCHARTS, TABLES OR GRAPHS?

Deciding between a photograph, a sketch, a flowchart, a graph or a table depends on whether the author intends to communicate a point at a glance, show trends and relationships or show the exact data. As a guide, use:

- A *photograph* or a *sketch* to communicate a point at a glance.
- A *flowchart* to describe a series of events that culminated into a single goal.
- A *graph* to show trends and relationships.
- And a *table* to show exact numbers/data as well as give plenty of information in words, *e.g.*, when summarizing a long list of previous research like in Binks and Tyowua (2016).

The style of presenting photographs, sketches, flowcharts, graphs and tables vary from journal to journal so always consult the *Guide for Authors* or check recent articles of the target journal for clues on recommended presentation styles. However, all journals prescribe "stand-alone" photographs, sketches, flowcharts, graphs and tables. "Stand-alone" means they should be understood without reference to the body of the paper. So, ensure that your photographs, sketches, flowcharts, graphs and tables as well as their caption contain all the necessary information required for making them understandable without reference to the body of the paper. The goal of the subsequent subsections is to give you the basic rules for presenting your results in either photographs, sketches, flowcharts, graphs or tables.

DOI: 10.1201/9781003186748-11

11.3   PREPARATION OF PHOTOGRAPHS

As used here, photographs refer to images from digital cameras and microscope images. Photographs communicate a point at a glance and may also prove a finding. When used, photographs give the reader the opportunity to also see what the author observed during the experiment. Use photographs of high resolution (at least 300 dots per inch) and do not over edit them to avoid result falsification. There are various software applications for editing photographs, but I recommend Microsoft PowerPoint because it is easy to use and has all the basic tools (cropping, labeling, image size minimization or maximization) required for editing a photograph and will save you the purchasing cost of other photo-editing software applications. As a guide:

- Be sure that the point you intend to communicate using the photograph is clearly visible.
- When adjusting the brightness, balance and/or contrast of a photograph, be sure that the basic features you intend to communicate using the photograph remain visible. When multiple photographs are involved, apply the same adjustments to all of them – it is wrong to adjust some images and leave out others (Figure 11.1).
- When grouping separate photographs together, differentiate them with a line and clear labels. For example, the images in Figure 11.1 are groupings of separate images. The images are clearly labeled (a) and (b), and they are separated by lines compared with Figure 11.2 where the images are not separated.

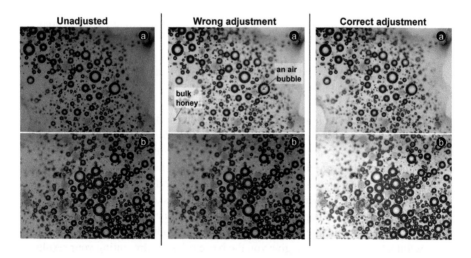

FIGURE 11.1 Optical microscope images of honey foams: (left) the brightness of the images is unadjusted, (middle) the brightness of image (a) has been adjusted but that of (b) remains unadjusted and (right) the brightness of both images (a) and (b) has been adjusted. Adjusting the brightness of one image and leaving the other is wrong while adjusting both images is correct.

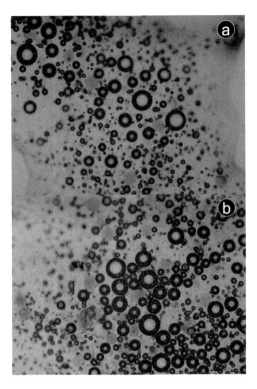

FIGURE 11.2 The microscope images (a) and (b), in Figure 11.1, are grouped without a separating line. Notice how it is difficult to know which image is which without the separating line?

- When increasing or decreasing the size of a photograph, be sure that its length and width change proportionately. For example, compare the correct re-sizing of the microscope image in Figure 11.1a, where the length and the width change proportionately, with re-sizing it incorrectly, where the length and the width change disproportionately (Figure 11.3). The spherical shape of the air bubbles in the microscope image is altered when the length and the width change disproportionately (Figure 11.3).
- Place a scale bar on photographs, especially microscope images, to help readers estimate size in relation to the photograph. For example, it is impossible to estimate the size of air bubbles on the microscope images of the foams shown in Figures 11.1–11.3 without the scale bar on them. A scale bar (Figure 11.4) is either a line or a bar used to indicate the size (length) of features and the distances between them on an image.

11.4   DRAWING SKETCHES AND FLOWCHARTS

A sketch is a rough drawing or diagram (to scale or not to scale) of something, while a flowchart is a diagram that shows the sequence of a process or an event. Sketches and

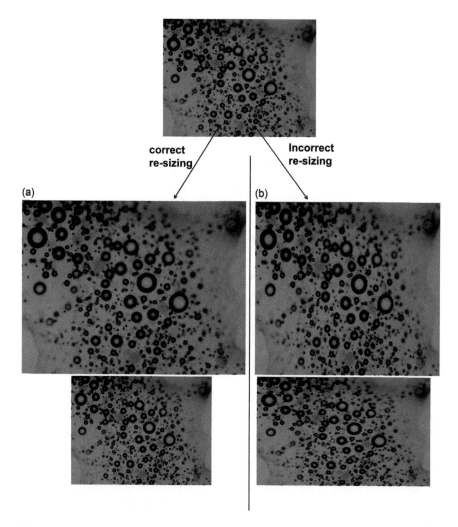

FIGURE 11.3 Correct (a) *versus* incorrect (b) re-sizing of an image. The length and the width of the image are changed proportionately in the former (a), but they are changed disproportionately in the latter (b).

flowcharts play a central role in scientific communication (Rowley-Jolivet 2002) – in fact, seven out of ten scientific papers contain a sketch and/or a flowchart. In the absence of photographs, sketches are used to give the reader a basic visual idea of an observation. For example, the microscope images of the honey foams shown in Figures 11.1–11.4 can be sketched (Figure 11.5). The sketch gives the reader an idea of the key microscopic features of the foams: the air bubbles are spherical, with variable size and with some air bubbles sticking onto others. Sketches are also used to communicate complex ideas (Rowley-Jolivet 2002).

Flowcharts are used in scientific writings to give a step-by-step summary of an experimental procedure or the working principle of an equipment. In some cases, sketches are incorporated into flowcharts. For example, Lutier and Vaissière (1993)

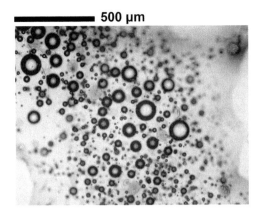

FIGURE 11.4   An optical microscope image of a honey foam with a scale bar of 500 µm.

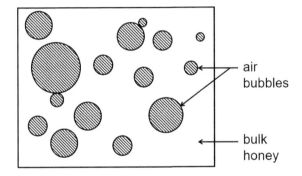

FIGURE 11.5   A sketch (not to scale) to illustrate the microscope images of the honey foams shown in Figure 11.1. Just like in Figure 11.1, the sketch shows spherical air bubbles of different sizes (with some sticking onto others) dispersed in honey.

used a flowchart (Figure 11.6) to summarize the extraction and analysis of pollen grains in honey. The key steps in the procedure – filtration, acetolysis, centrifugation, storage and mounting – are given in the flowchart with minor additional details. As a guide:

- Keep flowcharts clear and simple like in Figure 11.6.
- Use boxes for each entry, *i.e.*, step.
- Keep text in flowcharts to a minimum.
- Use color only if it will make the flowchart more understandable.
- Use software packages that are acceptable by the journal to draw your flowchart. Many journals accept CorelDraw, Microsoft PowerPoint, WordPerfect, Adobe Illustrator and Macromedia Freehand.
- And save the flowchart in the format (jpeg, jpg, tiff, gif, *etc.*) recommended by the journal.

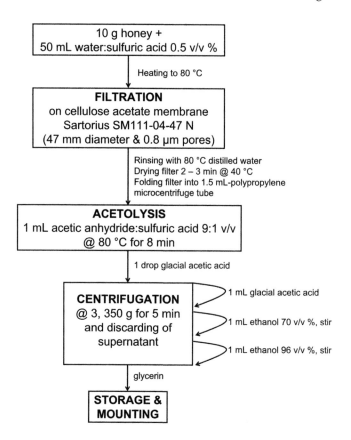

FIGURE 11.6   Flowchart for the extraction and analysis of pollen in honey. The flowchart shows sequence of steps, namely filtration, acetolysis, centrifugation, storage and mounting, necessary for extracting and analyzing pollen in honey. Where necessary, additional text is added for clarity. From Lutier and Vaissière (1993, *Rev. Palaeobot. Palyno.*, **78**, 129–144).

11.5   DRAWING GRAPHS

Graphs are central in scientific papers because they are very efficient for presenting complex data (Smith et al. 2000, Schofield 2002). That is why it is difficult to find a scientific paper without a graph. Graphs are used to show trends and relationships (*i.e.*, cause-and-effect) and make comparison at a glance (Ng and Peh 2009). To take advantage of the power of graphs, ask yourself whether or not it is necessary to transform your data into a graphical representation. If your answer is "No", present the data in a suitable table. If it is "Yes":

- Decide whether your graph is meant to show a trend, a relationship or make comparison and design the graph around your need.
- Decide the type of graph your data is more suited for – line graph (more suitable for showing trends), bar graph (more suitable for comparing magnitudes) and pie chart (more suitable for showing proportions of a whole). For line and bar graphs, choose the appropriate axes – normally the independent

variable is on the abscissa while the dependent variable is on the ordinate axis (Annesley 2010). For example, if "*A*" was varied to obtain "*B*", "*A*" is an independent variable while "*B*" is a dependent variable.

- Decide the appropriate software package to use. This may be difficult given the number of software packages available for plotting graphs, but Microsoft Excel spread sheet is commonly used.
- Do not overload the graph with data and use a suitable scale so that the graph will be simple, clear and easily understandable. Aim to have less than *five* trend lines per line graph and leave adequate spaces between columns in bar charts. Draw the bar chart vertically if it has less than *ten* grouped columns (Figure 11.7) and horizontally if it has more than *ten* grouped columns (Figure 11.8, from Binks et al. 2014).
- Use a combination of solid and dotted lines or different marker types (●, ○, ◇, ◆, △ or ▲) to differentiate data sets on the same line graph. Similarly, use different color or pattern fills (hatched, crosses or dots) to differentiate between data sets for bar graphs (see Figures 11.7 and 11.8).
- Know that default settings do not often give high-quality graphs; therefore, edit graphs before exporting them into your manuscript. Aim to edit your graph to font type *Times New Roman*, font size *12–14 pt*, marker size (*i.e.*, plotted data point) *5 pt* and line joining data points *1 pt*. Additionally, remove all grid lines (Green 2006), but enclose the graph with four lines (1 pt). Also, subdivide the axes equally with inner ticks. An example of an edited graph (based on these specifications), plotted in Microsoft Excel spread sheet, is shown in Figure 11.9 (Tyowua and Binks 2020). Always aim to make your graphs look like Figure 11.9 irrespective of the software used.

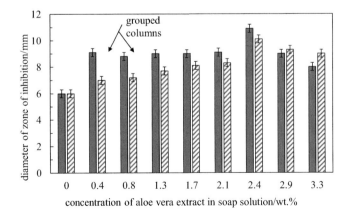

FIGURE 11.7   Plot of average diameter of zone of inhibition for *Staphylococcus aureus* (■) and *Pseudomonas aeruginosa* (▨) *versus* concentration of *Aloe vera* extract in soap solution. The error bars are standard deviation of three measurements.

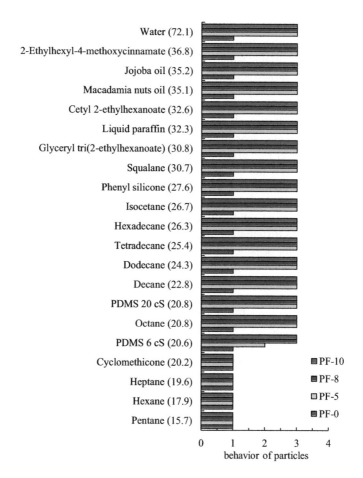

FIGURE 11.8 Behavior of PF-sericite particles (50 mg) on the surface of various liquids (3 mL) at rest. Abscissa: 1 – particles were wetted by the liquid, 2 – particles were partially wetted by the liquid, 3 – particles were not wetted by the liquid. Ordinate: number in brackets is liquid surface tension (±0.1 mN m⁻¹) at 25°C. From Binks et al. (2014, *Soft Matter* **10**, 578–589).

11.6   DRAWING TABLES

Tables are used to present data that cannot be easily described using only words or to present "raw" data – *i.e.*, not processing them into graphs, bar charts or pie charts. In some cases, tables are also used to show interrelationship between data. Because tables are meant to support the text and not for duplicating the text or a graph (Kotz and Cals 2013), it is wrong to present a table and a graph, bar chart or pie chart with the same set of data. There are two types of tables, namely *informal* (also in text) and *formal* tables (Coghill and Garson 2006). Informal tables are less common in scientific writings than formal tables. Informal tables contain a minimum of *two* lines (or rows) and a maximum of *four* columns (each with or without a heading). An informal table is without title/caption, numbering as well as footnote, and it is placed in the text immediately after an introductory sentence. For example, the following reagents were used (introductory sentence for Table 11.1).

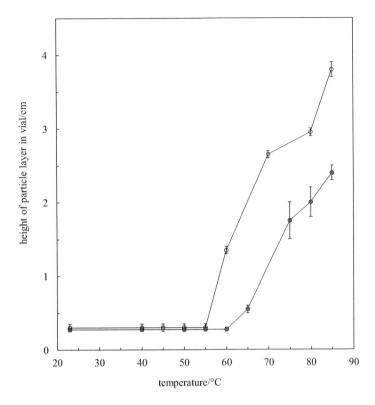

FIGURE 11.9   Plot of height of Expancel® micro-spherical plastic particle layer in a glass vial *versus* temperature in the absence (filled symbols) and presence (open symbols) of water. Above a certain critical temperature (≥55°C), the height of particle layer increases with increase in temperature. From Tyowua and Binks (2020, *J. Colloid Interface Sci.*, **561**, 127–135).

TABLE 11.1

An Example of an Informal Table

Reagent	Source	Purity/%
NaOH	BDH Chemicals Ltd	98
NH_4OH	May and Baker Ltd	25
HCl	May and Baker Ltd	36
HNO_3	BDH Chemicals Ltd	70
H_2SO_4	May and Baker Ltd	96
KI	Sigma-Aldrich	99.9
$CaCl_2$	BDH Chemicals Ltd	95
$Pb(NO_3)_2$	Pure Chemicals Ltd	99
$CuSO_4 \cdot 5H_2O$	BDH Chemicals Ltd	98
$FeCl_3 \cdot 6H_2O$	BDH Chemicals Ltd	98
$Na_2S_2O_3 \cdot 5H_2O$	BDH Chemicals Ltd	99
$AgNO_3$	BDH Chemicals Ltd	99.9
Na_2CO_3	BDH Chemicals Ltd	70.6

heading

columns

body

rows

TABLE 11.2

Emissions of Biodiesel for Some Model Diesel Vehicles

Vehicle	Fuel	Emissions (g km^{-1})		Hydrocarbon	Particulate Matter	Sulfur Oxides
		Nitrogen Oxides	CO			
Peugeot Partner	B100	2.05	9.37	0.54	2.68	0
Peugeot Partner	B20	1.86	17.73	1.32	4.71	0.004
Renault Kangoo	B100	2.23	9.22	0.49	3.06	0
Renault Kangoo	B20	1.92	17.36	1.26	5.63	0.003
Dacia Pickup	B100	2.15	9.42	0.56	2.59	0
Dacia Pickup	B20	1.91	18.29	1.35	4.63	0.005

Source: Demirbas (2009, *Appl. Energy*, **86**, 108–117)

Formal tables are more common in scientific writings than informal tables, and they contain at least *three* rows and at least *three* interrelated columns (Coghill and Garson 2006). This notwithstanding, some formal tables have two rows and/or two columns, but such tables are discouraged by publishers. Contrary to an informal table, a formal table has a title/caption, a number and a footnote in some cases. Unlike informal tables which are not cited in the text before being displayed, formal tables are cited in the text before they are displayed. An example is Table 11.2, showing the amount of biodiesel emission of some model of diesel vehicles (from Demirbas 2009).

The contents of a formal table is either *descriptive* (giving detailed information) or *declarative* (showing trend or relationship) (Gustavii 2017). Preparing a formal table is more challenging than preparing an informal table. As a guide:

- Aim to have a single table whenever possible rather than several of them; however, for the sake of printing, tables should have 3–13 columns so that they can fit well into a page when printed. Consider splitting tables larger than 13 columns. Generally, tables typically have more row entries or headings than column entries.
- When using Microsoft Office, always prepare tables in Microsoft Excel and export them into the Word document or use the table feature in Word.
- Optimize the use of space in your table and make them simple, concise and easily understandable.
- Because comparing numbers arranged down a column is easier than across a row, always arrange numbers down a column when making comparison.
- Use one type of alignment for column headings (left, right or center).
- For *declarative* contents, aim to have only one conclusion per table.
- The size of table entries should be 10–12 pt.
- Use footnotes to explain all the symbols used in the table.

- Use the leftmost column as the *stub* or *reading* column – *i.e.*, the reference point for all other columns. Indent sub-stub entries and be sure that both the stub and sub-stub entries are consistent with the text, logical and grammatically parallel throughout the table. For example, the "vehicle" column in Table 11.2 is the stub, and all other columns take reference from it.
- Use the *straddle rule* for a column whose heading applies to more than one column. The straddle rule is a horizontal line drawn above columns with the same heading, *e.g.*, the line between "emissions" and the columns containing entries of "nitrogen oxides", "CO", "hydrocarbons", "particulate matter" and "sulfur oxides" in Table 11.2 is a straddle rule.

11.7   WRITING CAPTIONS FOR FIGURES AND TABLES

A.   FIGURE CAPTION

In line with the requirement that figures should "stand alone" – *i.e.*, be understood without reference to the body of the paper, a figure caption contains (not in a sentence form):

- A *title*, which says what the figure is about.
- The *message* conveyed by the figure.
- An explanation of models or results given in the figure as well as method used.
- An explanation of symbols used.
- An explanation of units and any statistical symbols used.

Full sentences, with clear punctuations, are used for adding more information to figure captions. Depending on the figure, the caption may not contain all of these elements. For some figures, it is enough to state the message conveyed as the caption.

Activity 11.1 *Analysis of a Figure Caption*

Identify the *title*, an explanation of *general* and *statistical symbols* used in the following figure caption taken from a journal paper. Notice that the caption does not contain all the elements stated earlier.

FIGURE CAPTION

"Responses of corn and soybean grain yields to chemical fertilizer and various swine manures applied to a Brookston clay loam soil from 2004 to 2011. LM, SM and MC represent liquid, solid and composted swine manure, respectively. Error bar is the standard error of the mean ($n = 12$). Different letters over the bars indicate significant differences at the $p \leq 0.05$ level".
From Hao et al. (2015, *Nutr. Cycl. Agroecosyst.*, **103**, 217–228)

B.   TABLE CAPTION

Similar to figures, tables are also required to "stand-alone"; therefore, a table caption normally contains (not in a sentence form):

- A *title*, which says what the table is about.
- The *message* conveyed by the table (especially if the content is declarative, *i.e.*, showing a trend or a relationship).
- An explanation of all symbols used.
- An explanation of units and any statistical symbols used.

Similar to figures, complete sentences, with clear punctuations, are used for adding more information to table captions. Depending on the table, the caption may not contain all of these elements and for tables with declarative contents, stating the message conveyed by the table may be enough for the caption, *e.g.*, Table 11.3 (from Cho et al. 2005). I have given the table a declarative title for illustration's sake, but the original title of the table is descriptive, "effective diameters of bubbles generated by sonication in pure water" (Cho et al. 2005).

In addition to the caption, tables also contain footnotes with further explanation so as to stand-alone. There are *two* forms of footnotes, namely *general* and *specific* footnotes. A *general* footnote contains an SI unit that applies to all entries in the table, explanatory notes on abbreviations or symbols that apply to the entire table, explicit experimental conditions (if not given or different in the text) and general sources of data. A *specific* footnote contains units that are too large to fit into a column heading and explanatory notes on abbreviations or symbols, experimental details, statistical significance of entries as well as sources of data that apply to specific columns in the table. Depending on the journal, footnotes are indicated by superscripts using *, †, ‡, §, ¶, ‖, *a*, *b* or *c* or labeled as "Note" or "Source". For some journals, the footnote is placed below the table without any superscripts or labels (*e.g.*, Table 11.4 from Idris et al. 2011). Use the *Guide for Authors* to know the recommended style of the journal.

11.8   PRESENTING FIGURES AND TABLES

Figures and tables are numbered sequentially and presented sequentially – *i.e.*, if figures are numbered as Figures 1, 2, 3, …, "Figure 1" must be presented first, followed by "Figure 2" and then "Figure 3" …. Similarly, for a list of tables say Tables 1, 2, 3, …, "Table 1" is presented first, followed by "Table 2" and then "Table 3".… Generally:

TABLE 11.3

Increase in Effective Bubble Diameter with Water Sonication Time

Time (min)	Effective Diameter (nm)
0	749
10	745
20	749
30	753
40	774
50	796

Source: Cho et al. (2005 *Colloids Surf. A*, **269**, 28–34)

TABLE 11.4

Chemical Properties of Sudanese Honeys

Honey Source	% Moisture	% Ash	% Nitrogen	% Protein
Cucurbita maxima Duch. (Pumpkin)	16.60 ± 0.61^c	0.12 ± 0.01^b	0.035	0.22 ± 0.01
Acacia nilotica (Sunut)	16.20 ± 0.55^c	0.13 ± 0.01^b	0.033	0.21 ± 0.01
Ziziphus spina-christi (Sidir-mountain)	17.40 ± 0.72^b	$0.12 \pm 0.01b$	0.037	0.23 ± 0.01
Balanites aegyptiaca (Heglig)	17.47 ± 0.81^b	0.13 ± 0.01^b	0.037	0.23 ± 0.01
Ziziphus spina-christi (Sidir)	17.77 ± 0.60^b	0.13 ± 0.01^b	0.032	0.20 ± 0.01
Acacia seyal (Talih)	16.70 ± 0.54^c	0.16 ± 0.01^b	0.033	0.21 ± 0.02
Azadirachta indica (Neem)	21.27 ± 0.35^a	1.21 ± 0.1^a	0.046	0.29 ± 0.02

All determinations were carried out in triplicate and mean value ± standard deviation reported. Mean values in the same column having different superscript letters differ significantly ($p > 0.05$). From Idris et al. (2011, *Int. J. Food Prop.*, **14**, 450–458)

- Irrespective of their position in a sentence, numbered figures and tables begin with capital letters – *i.e.*, "Figure 1" or "Table 1", not "figure 1" or "table 1". In some journals, "figure" is shortened to "fig.", *e.g.*, "Fig. 1" for "Figure 1".
- Parts of a figure are designated by alphabets (lower or upper case) or Arabic numerals, *e.g.*, Figure 1a or 1A, Figure 1(a) or Figure 1(A) and Figure 1-I or Figure 1(i).

Figures and tables are *cited* or *called-out*, *i.e.*, mentioned in the text, before they are presented. Figures and tables are suitably called-out either at the beginning, in the middle or at the end of a sentence.

- *Beginning of a sentence*: "Figure 4 shows SEM images of particle sizes formed in light, medium and dark honeys" (Brudzynski et al. 2017). "Table 1 presents some physicochemical parameters (moisture content, °Brix concentration, fructose, glucose, fructose/glucose ratio) of the seven honeys analyzed" (Oroian 2013).
- *Middle of a sentence*: "The plot of bubble size against particle concentration (Figure 5) shows an initial increase with a maximum at 7 wt. % and a decrease thereafter" (Tyowua and Binks 2020). "The bubble size in the foams increases with increasing temperature (Figure 10), reaching a maximum at 60 °C with a slight decrease at 70 °C" (Tyowua and Binks 2020).

- *End of a sentence*: "With further dilution, the rate of decrease in the effective diameter of particles became essentially superimposable, showing that molecular crowding is needed for the formation of macromolecular superstructures (Figure 3)" (Brudzynski et al. 2017). "There were no significant differences between tested honeys with regard to these parameters (Table 1)" (Brudzynski et al. 2017). "We illustrate this in Figure 1" (Hawking and Hertog 2018).

Multiple figures and tables are similarly called-out at the beginning, in the middle or at the end of a sentence. Examples include: "Figures 1–5 illustrate HPLC chromatogram of the sugar analysis of honey samples in different concentrations" (El Sohaimy et al. 2015), "typical signals of CO_3^{2-} are observed in fixed spots of the mapping spectra in (Figure 2 c, d), 400 different positions (spot size $3 \times 3\,\mu m^2$) in one $50 \times 50\,\mu m^2$ area were selected for the Raman mapping" (Li et al. 2017), "downregulation of $HIF2\alpha$ did not affect cell growth in vitro, but was sufficient to impair tumor growth in-vivo (Figure 2B–2D)" (Kondo et al. 2003).

11.9 CREDIT LINE IN FIGURE AND TABLE CAPTIONS

A figure or a table *Credit line* refers to wordings that acknowledge their original source. The credit line usually follows the figure caption, perhaps in parenthesis or table footnote. With the exception of public domain materials, *e.g.*, US government work, which can be reproduced without formal written permission, reproducing a previously published figure or table (in part or whole) requires written permission of the copyright owner (usually the publisher) even if the figure or table was published by the author. In some cases, charges apply and the permission license is issued only after payment has been made. A copy of the permission license is submitted along with the manuscript, and the caption of the figure or table footnote ends with a credit line. Contrarily, permission is not needed when using previously published data or text to construct an original figure or table; however, the source of the data or text must be referenced. Work published under *Creative Commons* license, designated as "CC-BY" or "CC BY-NC", are partially protected by copyright and contain elements that can be used without formal permission, but they may not be in public domain. Always find out what elements of the work require or do not require formal permission and use them accordingly. The wordings of a credit line depend on the publisher and are often given in the permission license. Always look out for the acceptable format in the permission license. Nonetheless, typical wordings of a credit line for reproduced materials include:

- "Reprinted/Reproduced with permission from ref. XX. Copyright year Copyright Holder's name" (*most publishers*). For example, "Reprinted with permission from ref. 11. Copyright 2015 Oxford University Press".
- "Reprinted/Reproduced from ref. XX. Copyright year American Chemical Society" (*American Chemical Society*), *e.g.*, "Reprinted from ref. 18. Copyright 2015 American Chemical Society".
- "Reprinted/Reproduced with permission from Author Names (Year of Publication). Copyright Year Copyright Owner's Name", *e.g.*, "Reprinted with permission from Moore (2013). Copyright 2013 American Institute of Physics".

- "Reprinted/Reproduced from Author Names (Year of Publication). Copyright Year American Chemical Society", *e.g.*, "Reprinted from Moore (2013). Copyright 20013 American Chemical Society".
- "Reprinted/Reproduced from ref. XX with permission from The Royal Society of Chemistry", *e.g.*, "Reproduced from ref. 21 with permission from The Royal Society of Chemistry".
- "From Author Name(s), Name of Journal, Volume, Pages, Year" (*Taylor and Francis*), *e.g.*, "Demirbas, A., *Appl. Energy*, 86, 108–117, 2009".
- "Quoted from ref. XX, Copyright, Copyright Owner's Name", *e.g.*, "Quoted from ref. 21, Copyright, Wiley".

The credit line for using or adapting part of a figure or a table is similar to that for reproducing the entire figure or table; however, "Reprinted"/ "Reproduced" is replaced with "Adapted". An example is "Adapted from ref. 21 with permission from the Royal Society of Chemistry".

Activity 11.2 *Identification of Credit Lines in a Figure Caption*

Identify the credit lines in Figure 11.10, taken from Tyowua et al. (2019).

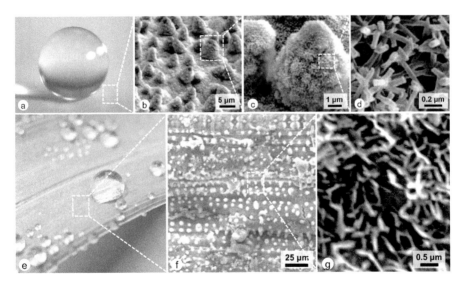

FIGURE 11.10   (a) Photograph of a quasi-spherical water drop on the surface of a Lotus leaf. (b)–(d) SEM micrographs of the Lotus leaf surface, at increasing magnifications, showing its rough hierarchical microstructure. Reprinted from Bhushan et al. 2009, with permission from the Royal Society Publishing, London. (e) Photograph of quasi-spherical water drops on the surface of a rice leaf. Reprinted from Shin et al. 2016, under the Creative Common Attribution License of MDPI. (f) and (g) SEM micrographs of the rice leaf at increasing magnifications, showing its rough hierarchical microstructure. Reprinted from Guo et al. 2007, with permission from Elsevier". From Tyowua et al. (2019, *Rev. Adhesion Adhesives* 7(2):195–231).

FURTHER READING

Nicol, A.A., and P.M. Pexman. 2003. *Displaying Your Findings: A Practical Guide for Creating Figures, Posters, and Presentations*: American Psychological Association, Washington.
* Contains excellent practical guidelines for creating and displaying graphs, photographs, sketches and charts.
Coghill, A.M., and L.R. Garson. 2006. *The ACS Style Guide: Effective Communication of Scientific Information*: Oxford University Press, New York.
* Contains useful suggestions for preparing and presenting figures, tables as well as credit line in scientific research papers.
Parija, S.C., and V. Kate, eds. 2017. *Writing and Publishing a Scientific Research Paper*: Springer, Singapore.
* The presentation and uses of figures (graphs, sketches and photographs) and tables in the results section of scientific research papers are discussed.
Rougier, N.P., M. Droettboom, and P.E. Bourne. 2014. Ten simple rules for better figures. *PLoS Comput. Biol.* 10 (9):e1003833.
* Contains excellent suggestions for creating figures for scientific papers. A list of possible software applications for creating figures is also given.
Weissberg, R., and S. Buker. 1990. *Writing Up Research*: Prentice Hall Englewood Cliffs, NJ.
* Discusses essential elements of the results section of a research paper and gives useful suggestions for writing the results section. Suggestions for writing other parts of a research paper are also given.

REFERENCES

Annesley, T.M. 2010. Put Your Best Figure Forward: Line Graphs and Scattergrams. *Clin. Chem.* 56 (8):1229–1233.
Binks, B.P., T. Sekine, and A.T. Tyowua. 2014. Dry oil powders and oil foams stabilised by fluorinated clay platelet particles. *Soft Matter* 10 (4):578–589.
Binks, B.P., and A.T. Tyowua. 2016. Oil-in-oil emulsions stabilised solely by solid particles. *Soft Matter* 12 (3):876–887.
Brudzynski, K., D. Miotto, L. Kim, C. Sjaarda, L. Maldonado-Alvarez, and H. Fukś. 2017. Active macromolecules of honey form colloidal particles essential for honey antibacterial activity and hydrogen peroxide production. *Sci. Rep.* 7 (1):7637.
Bhushan, B., Y.C. Jung, and K. Koch. 2009. Micro-, nano- and hierarchical structures for superhydrophobicity, self-cleaning and low adhesion. *Phil. Trans. R. Soc. A* 367:1631–1672.
Cho, S.-H., J.-Y. Kim, J.-H. Chun, and J.-D. Kim. 2005. Ultrasonic formation of nanobubbles and their zeta-potentials in aqueous electrolyte and surfactant solutions. *Colloids Surf. A* 269 (1):28–34.
Coghill, A.M., and L.R. Garson. 2006. *The ACS Style Guide: Effective Communication of Scientific Information*: Oxford University Press, New York.
Demirbas, A. 2009. Political, economic and environmental impacts of biofuels: A review. *Appl. Energy* 86:S108–S117.
El Sohaimy, S.A., S.H.D. Masry, and M.G. Shehata. 2015. Physicochemical characteristics of honey from different origins. *Ann. Agric. Sci.* 60 (2):279–287.
Green, J. 2006. Graphs. *Chest* 130 (2):620–621.
Gustavii, B. 2017. *How to Write and Illustrate a Scientific Paper*: Cambridge University Press.
Guo, Z., and W. Liu. 2007. Biomimic from the superhydrophobic plant leaves in nature: Binary structure and unitary structure. *Plant Sci.* 172 (6):1103–1112.
Hao, X.J., T.Q. Zhang, C.S. Tan, T. Welacky, Y.T. Wang, D. Lawrence, and J.P. Hong. 2015. Crop yield and phosphorus uptake as affected by phosphorus-based swine manure application under long-term corn-soybean rotation. *Nutr. Cycl. Agroecosyst.* 103 (2):217–228.

Hawking, S.W., and T. Hertog. 2018. A smooth exit from eternal inflation? *J. High Energ Phys.* 2018 (4):147.

Idris, Y.M.A., A.A. Mariod, and S.I. Hamad. 2011. Physicochemical properties, phenolic contents and antioxidant activity of Sudanese honey. *Int. J. Food Prop.* 14 (2):450–458.

Kondo, K., W.Y. Kim, M. Lechpammer, and W.G. Kaelin Jr. 2003. Inhibition of HIF2α is sufficient to suppress pVHL-defective tumor growth. *PLoS Biology* 1 (3): e83.

Kotz, D., and J.W. Cals. 2013. Effective writing and publishing scientific papers, part VII: tables and figures. *J. Clin. Epidemiol.* 66 (11):1197.

Li, Y., B. Xu, H. Xu, H. Duan, X. Lü, S. Xin, W. Zhou, L. Xue, G. Fu, A. Manthiram, and J.B. Goodenough. 2017. Hybrid polymer/garnet electrolyte with a small interfacial resistance for lithium-ion batteries. *Angew. Chem. Int. Ed.* 56 (3):753–756.

Lutier, P.M., and B.E. Vaissière. 1993. An improved method for pollen analysis of honey. *Rev. Palaeobot. Palyno.* 78 (1):129–144.

Ng, K.H., and W.C. Peh. 2009. Preparing effective illustrations. Part 1: Graphs. *Singap. Med. Journal* 50 (3):245–249.

Oroian, M. 2013. Measurement, prediction and correlation of density, viscosity, surface tension and ultrasonic velocity of different honey types at different temperatures. *J. Food Eng.* 119 (1):167–172.

Rowley-Jolivet, E. 2002. Visual discourse in scientific conference papers A genre-based study. *English for Specific Purposes* 21 (1):19–40.

Schofield, E.K. 2002. Quality of graphs in scientific journals: An exploratory study. *Science Editor* 25 (2):39–41.

Shin, S., J. Seo, H. Han, S. Kang, H. Kim, and T. Lee. 2016. Bio-inspired extreme wetting surfaces for biomedical applications. *Materials* 9 (2):1–26.

Smith, L.D., L.A. Best, D.A. Stubbs, J. Johnston, and A.B. Archibald. 2000. Scientific graphs and the hierarchy of the sciences: A latourian survey of inscription practices. *Soc. Stud. Sci.* 30 (1):73–94.

Tyowua, A.T., and B.P. Binks. 2020. Growing a particle-stabilized aqueous foam. *J. Colloid Interface Sci.* 561: 127–135.

Tyowua, A.T., M. Targema, and E.E. Ubuo. 2019. Non-wettable surfaces–from natural to artificial and applications: A critical review. *Rev. Adhesion Adhesives* 7 (2):195–231.

12 Discussion Section

12.1 FUNCTIONS OF THE DISCUSSION SECTION

The results and the discussion sections are merged together in some journals where the results are presented and discussed concurrently, perhaps to avoid repetition of the contents of the results section in the discussion section. Whether the results and the discussion sections are merged together or not, the primary function of the discussion section is to explain the meaning of the results to the reader (Hess 2004). On this basis, the discussion section: (i) evaluates the results critically *vis-à-vis* the aim and the hypothesis of the work, (ii) interprets and gives meaning to the results based on the evaluation made, (iii) relates the work to previous ones and (iv) explains how the work advances the knowledge base of the area.

12.2 WRITING THE DISCUSSION SECTION – TELLING THE STORY

- Writing a research paper, from the title to the conclusion, is akin to story-telling. The discussion section is where most of the story is told; therefore, it must be well-crafted to contain every information the reader will typically be looking out for.
- Firstly, identify the central message contained in the overall results and structure your discussion (story) around it. Results from individual experiments may have their own key message, but they must be linked to the central message contained in the overall results. The central message must be in turn linked to the hypothesis of the work.
- Secondly, organize the discussion section into subsections or paragraphs that relate directly to individual experiments performed in the work – *i.e.*, the results from each experiment should be discussed in separate subsections or paragraphs as the case may be. However, connect these subsections or paragraphs and show how they support the central message of the paper.
- The structure of the discussion section can be likened to a *cone* (Figure 12.1) because it starts narrowly with the results, widening gradually as existing results are compared and generalizations are made (Cals and Kotz 2013).

To obtain the cone structure, organize each subsection or paragraph according to the following moves (Hess 2004, Docherty and Smith 1999): (i) give a *one-sentence* summary of the objective of the experiment; (ii) give a *one-sentence* summary of the experimental procedure; (iii) show or locate the results of the experiment; (iv) evaluate the results, stating whether they were expected or not; (v) interpret the results, stating any ramifications and importance; (vi) relate the results to those from similar

DOI: 10.1201/9781003186748-12

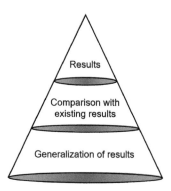

FIGURE 12.1   Schematic illustration of the cone-shaped nature of the discussion section of a scientific research paper. The results occupy the upper part of the cone. In the middle of the cone, the scope of the results is widened by comparing them with existing published results. At the base of the cone, the scope of the results is widened further as generalizations are made.

previous studies; and finally (vii) acknowledge any limitations and recommend further research. Move to the next set of results and follow the same structure. Discuss the entire results using this structure. Although one or more of these components are absent in the discussion sections of some scientific papers like in Example 12.1, this structure is immediately obvious in majority of others, either in the sequence given here or in other sequences more comfortable to the author. For instance, the discussion subsection analyzed in Example 12.1 does not follow the sequence given here, but its structure has elements synonymous with those given here.

Example 12.1 *Analysis of a Discussion Subsection from Li et al. 2020*

Discussion subsection: Characteristics of Egg White Protein Microgel"
"The egg white protein formed a thermally crosslinked gel at 6.25 wt.% protein (Supplementary Fig. S1) which was then used to fabricate the microgel particles via the jet homogenization process. As can be seen from Fig. 1a, egg white protein microgels had a mean hydrodynamic diameter (D_h) of ~359 nm, with a particle size distribution showing the most prominent peak in the region 100–1000 nm and a relatively small peak below 100 nm. The smaller peak probably represents egg white protein that somehow escaped the microgel formation process. Similar small peak has been previously observed in the case of whey protein microgel formation".

From Li et al. (2020, *Food Hydrocoll.*, **98**, 105292)

ANALYSIS

SHOWING RESULT AND GIVING SUMMARY OF EXPERIMENTAL PROCEDURE

The egg white protein formed a thermally crosslinked gel at 6.25 wt.% protein (Supplementary Fig. S1), which was then used to fabricate the microgel particles via the jet homogenization process.

SHOWING AND EXPLAINING RESULT

As can be seen from Fig. 1a, egg white protein microgels had a mean hydrodynamic diameter (D_h) of ~359 nm, with a particle size distribution showing the most prominent peak in the region 100–1000 nm and a relatively small peak below 100 nm.

INTERPRETING RESULT

The smaller peak probably represents egg white protein that somehow escaped the microgel formation process.

RELATING RESULT TO A SIMILAR ONE IN THE LITERATURE

Similar small peak has been previously observed in the case of whey protein microgel formation.

From Li et al. (2020, *Food Hydrocoll.*, **98**, 105292.)

Use *past tense* to state your results (findings) and *present tense* to interpret and/or explain the results. Returning to Example 12.1, "… egg white protein microgels had a mean hydrodynamic diameter (D_h) of ~359 nm, with a particle size distribution showing the most prominent peak in the region 100–1000 nm and a relatively small peak below 100 nm" is *past tense*, and it was used to the state finding. "The smaller peak probably represents egg white protein that somehow escaped the microgel formation process" is *simple present tense*, and it was used to interpret the result.

Activity 12.1 *Analyzing a Discussion Subsection*

Identify the various moves and sentences (past or present tense) used by the authors in the following discussion subsection

Discussion subsection: Effect of heat on particles in the absence and presence of water"

"According to the manufacturer, the micro-spherical particles expand when heated and so this was investigated. A photograph of particle samples heated alone and in the presence of water at various temperatures is shown in Fig. S1. The height of the particle layer is plotted against temperature for the different samples in Fig. S2. In the absence of water, the particles were sensitive to temperatures > 60 °C, leading to an increase in the height of the particle layer which increases with increasing temperature. With water, particle sensitivity occurred at a slightly lower temperature (> 55 °C), leading to water absorption and a five-fold increase in the height of the particle layer. The shift in temperature might be due to a specific interaction between the particles and water. By 85 °C, a 12-fold increment was obtained. Clearly, a higher particle layer is obtained when the particles are heated in the presence of water compared with when they are heated alone. These temperatures are lower than that reported by the manufacturer (80–95 °C) at which particle expansion begins".

From Tyowua and Binks (2020, *J. Colloid Interface Sci.*, **561**, 127–135)

12.3   DETAILED ELEMENTS OF THE DISCUSSION SECTION

i. **Summary of experimental objective**

In one sentence remind the reader of the objective of the experiment, stating any hypothesis, prediction or assumption. This sentence is the *topic sentence* of the subsection or paragraph and signals the contents of subsequent sentences. For example, "according to the manufacturer, the micro-spherical particles expand when heated and so this was investigated" (Activity 12.1). This sentence reminds the reader of the objective of the experiment. It also makes the reader to look forward to the results of the experiment.

ii. **Summary of experimental procedure**

Sometimes, authors summarize the experimental objective but leave out the experimental procedure or summarize the experimental procedure but leave out the experimental objective (*e.g.*, see Activity 12.1). When the experimental objective is omitted, the procedure becomes the topic sentence. Reminding readers of the procedure prevents them from returning to the experimental section for it, except for certain details not included in the summary.

iii. **Results**

After summarizing the objective and procedure of the experiment, show the reader the results which may be in the form of a figure or a table. Next, state the key findings (features, trends, relationships or statistical significance) embodied in the results. Sometimes, a single sentence is used to state the key findings and also locate the results, *e.g.*, "the egg white protein formed a thermally crosslinked gel at 6.25 wt.% protein (Supplementary Figure S1) which was …" (Example 12.1). It is common to find research papers with discussion subsections showing the results and stating the key findings straightaway without first summarizing the objective and procedure. This forces readers to return to the experimental section for both the objective and procedure, thereby slowing down their reading.

iv. **Evaluation of results**

Evaluate the results based on the objective of the experiment or aim of the work, stating any expectations and deviations. Evaluation and statement of a key finding can be done using a single sentence. An example is the sentence "regardless of the method, foam volume began to increase on raising the temperature to the onset of particle growth (60 °C) as *expected* and increased with increasing temperature" (from Tyowua and Binks 2020).

v. **Interpretation of results**

Explain the meaning as well as the importance of the results objectively. Keep the importance within the limit of your results, do not exaggerate. Also, state whether the results support your hypothesis or not. Your results may have other possible explanations – consider them even if they do not support your hypothesis because this will make your explanations more objective.

vi. **Relating results to previous studies**

Place your results side-by-side with those from similar published studies and bring out any similarities and/or differences to support your findings. Next, stress the new information contained in your results and explain how your findings have advanced the knowledge base of the area. You can also generalize your findings at this point, but this must be within the limit of your results.

vii. **Acknowledging limitations and suggesting further research**

Every work, including those in prestigious journals, has some limitations which sometimes become the basis for another piece of work (Hess 2004). Identify any limitations in your work and suggest further research. Your work may also give rise to unanswered questions which can form the basis for future research.

FURTHER READING

Glasman-Deal, H. 2020. *Science Research Writing: For Native and Non-Native Speakers of English*: Imperial College Press, London.
- Contains a comprehensive guide for writing the discussion section of a research paper. The discussion section of a research paper is discussed in terms of sentence structure and language style. With a substantial part of the book dedicated to grammar, the book is a useful material for improving grammar, especially with respect to scientific writing.

Weissberg, R., and S. Buker. 1990. *Writing Up Research*: Prentice Hall Englewood Cliffs, NJ.
- Discusses essential elements of the results and discussion section of a research paper and gives useful suggestions for writing it. Suggestions for writing other parts of a research paper are also given.

REFERENCES

Cals, J.W., and D. Kotz. 2013. Effective writing and publishing scientific papers, part VI: Discussion. *J. Clin. Epidemiol.* 66 (10):1064.

Docherty, M., and R. Smith. 1999. The case for structuring the discussion of scientific papers. *BMJ* 318 (7193):1224.

Hess, D.R. 2004. How to write an effective discussion. *Respirat. Care* 49 (10):1238.

Li, X., B.S. Murray, Y. Yang, and A. Sarkar. 2020. Egg white protein microgels as aqueous Pickering foam stabilizers: Bubble stability and interfacial properties. *Food Hydrocoll.* 98: 105292.

Tyowua, A.T., and B.P. Binks. 2020. Growing a particle-stabilized aqueous foam. *J. Colloid Interface Sci.* 561: 127–135.

13 Conclusion Section

13.1 FUNCTION OF THE CONCLUSION SECTION OF A RESEARCH PAPER

Some journals permit a separate conclusion section while others do not. In the absence of a separate conclusion section, the last paragraph in the discussion section is used for concluding the paper. The function of the conclusion section is to summarize the key findings. The author steps back and takes a wholistic look at the work and summarizes the key findings with respect to the aim or the hypothesis. This can be compared with the abstract section which summarizes the aim, methods, key findings and the main conclusion of the paper.

13.2 WRITING THE CONCLUSION SECTION

The conclusion section contains the same elements as the discussion section; thus, it can be considered as a "summarized version" of the discussion section of the paper. Therefore, just like the discussion section, the conclusion section of many scientific research papers typically:

- Reiterates the aim or the hypothesis of the work. The author relates the entire work back to the aim or the hypothesis, stating whether or not the aim has been achieved and also specifying the extent of achievement.
- Highlights key experiments just to remind the reader about what was done.
- Summarizes the key findings or the central message.
- Provides a possible explanation for the findings, *e.g.*, why something was observed rather than the other, a mechanism through which something happened, *etc.*
- States the contribution(s) of the work to the knowledge base of the area or the achievement(s) of the work.
- States the limitation of the work with respect to the subject area as well as other related areas.
- States potential applications or implications of the work.
- Suggests/recommends further research, perhaps from the limitation of the work, or an extension of the findings to other related areas which are not covered in the work.

Nonetheless, authors do not always follow this order, and sometimes, they marry some of the elements with others. Also, depending on the nature of the work reported, it is common to find conclusion sections without one or more of the aforementioned elements as illustrated in Example 13.1.

DOI: 10.1201/9781003186748-13

Example 13.1 *Analysis of the Conclusion Section of a Research Paper*

Analyze the conclusion section of the following research paper into the various elements listed in Section 13.2.

Title of Paper: Treatment of oil spills using organo-fly ash
Aim of Paper: This work aims at evaluating the possibility of removing oil and weathered oil from contaminated seawater using fly ash, a waste from thermal power plants, modified with hexadecyltrimethylammonium cation

CONCLUSION SECTION

"The study shows that hexadecyltrimethylammonium-fly ash obtained after modifying fly ash, a waste material, can be favorably deployed as a sorbent for oil spillage. The sorption behavior of fly ash was greatly enhanced when organically modified with hexadecyltrimethylammonium. Hexadecyltrimethylammonium-fly ash was also effective in removing dissolved organic carbon present in weathered oil contaminated seawater. This makes hexadecyltrimethylammonium-fly ash a better candidate compared to conventional sorbents. It was also observed that when the lighter oil fraction of the crude oil was increased, the sorption capacity was enhanced. This is also an advantage of this material in contrast to other sorbents like straw, which cannot retain large proportions of light oil. The predominant mode of sorption is probably the hydrophobic or nonpolar interaction between the long hydrocarbon chains of hexadecyltrimethylammonium with the oil".

From Banerjee et al. (2006, *Desalination* **195**, 32–39)

ANALYSIS

STATEMENT OF FULFILLMENT OF AIM

The study shows that hexadecyltrimethylammonium-fly ash obtained after modifying fly ash, a waste material, can be favorably deployed as a sorbent for oil spillage.

SUMMARY OF KEY FINDINGS

The sorption behavior of fly ash was greatly enhanced when organically modified with hexadecyltrimethylammonium. Hexadecyltrimethylammonium-fly ash was also effective in removing dissolved organic carbon present in weathered oil contaminated seawater.

ACHIEVEMENTS OF THE WORK

This makes hexadecyltrimethylammonium-fly ash a better candidate compared to conventional sorbents. It was also observed that when the lighter oil fraction of the crude oil was increased, the sorption capacity was enhanced. This is also an advantage of this material in contrast to other sorbents like straw, which cannot retain large proportions of light oil.

EXPLANATION OF FINDING

The predominant mode of sorption is probably the hydrophobic or nonpolar interaction between the long hydrocarbon chains of hexadecyltrimethylammonium with the oil.

Just like the abstract, verb tenses in the conclusion section depend on the element being written. For example, authors write the aim or hypothesis either in the *simple present* (see Example 13.1), *simple past* (*e.g.*, "in this work, the wetting characteristics that control the deposition pattern are explored" (Li et al. 2013)) or *present perfect tense* (*e.g.*, "the effect of the laser diameter, laser power density and exposure time on fluid flows, evaporation time and resultant distribution of suspended nanoparticles in evaporating droplets has been demonstrated" (Ta et al. 2016)). Authors use the *simple past tense* to highlight an experimental procedure, *e.g.*, "the hydrophobicity of the particles was confirmed by contact angle measurement and their ability to form stable aqueous foams and liquid marbles" (Tyowua and Binks 2021). Authors summarize key findings using the *simple present tense*, *e.g.*, "due to the localized heating, laser-induced flows drive particles to move and accumulate in any chosen area (within the droplet) with a selective pattern size" (Ta et al. 2016). Findings are also explained using the *simple present tense* (see Example 13.1). The *simple present tense* is also used to state the potential applications of the findings and also state the achievements of the work. For example, "this effect has potential applications in biotechnology and disordered photonic devices where high particle density and minimum deposition space are highly important" (potential application, Ta et al. 2016) and "unlike an earlier paper, we show how sheet-like structural materials form upon evaporating the continuous phase of a water-in-oil Pickering emulsion" (achievement, Tyowua and Binks 2021). Similarly, authors use the simple present tense to state the limit of their findings, *e.g.*, "measuring the oleogel bulk elastic properties that are relevant to prevent bubble dissolution is however challenging due to the thixotropic nature of the oleogel used in our experiments" (Saha et al. 2020). In order to suggest further research or an extension of the findings to other related areas which are not covered in the work, the simple present tense is used in combination with a modal auxiliary verb, *e.g.*, "can", "could", "should", "would", *etc.* For example, "although precipitated $CaCO_3$ particles are used here, other edible colloidal particles can be employed to make these emulsions" (extension of findings/recommendation, simple present tense + modal auxiliary verb "can", Tyowua et al. 2022).

Activity 13.1 *Analyzing the Conclusion Section of a Research Paper*

Analyze the conclusion section of the following research paper into the various elements listed in Section 13.2 as well as the various verb tenses used.

"Density functional theory has been used to elucidate the rational mechanistic approaches to Pd(0) generation from pincer palladacycles for catalysis in the Suzuki-Miyaura cross-coupling reactions of aryl bromides and phenyl-boronic acid. It was found that the presence of the base in the pre-catalyst activation is to significantly lower the activation energy barrier of the transmetallation step and overall reaction energy of the process. In addition, solvent effects are dependent on the mechanism in question, with or without base. Furthermore, the slow, controlled release of the "true, active catalyst" for reaction in the Suzuki-Miyaura coupling is more beneficial for the reaction and may be achieved either by a high activation energy barrier for the transmetallation step, or overall reaction energy of the pre-catalyst activation process, or both. On this backdrop, a good pre-catalyst would be one that has donor atoms that provide the large transmetallation, reductive elimination or overall reaction energies to control the release of catalytically active Pd(0). Finally, our investigations indicate that there is no significant difference in the catalytic activity of unsymmetrical and symmetrical pincer palladacycles in the Suzuki-Miyaura couplings investigated".

From Boonseng et al. (2017, *J. Organomet. Chem.* **845**, 71–81)

Activity 13.2 *Writing a Conclusion for an Experimental Report*

Using the information garnered in the chapter, write a conclusion for any completed simple experiment of your choice. This could be a routine or an occasional experiment in your laboratory.

FURTHER READING

Glasman-Deal, H. 2020. *Science Research Writing: For Native and Non-Native Speakers of English*: Imperial College Press, London.
  - Contains a comprehensive guide for writing the discussion and the conclusion sections of a research paper. These sections are discussed in terms of sentence structure and language style.
Parija, S.C., and V. Kate, eds. 2017. *Writing and Publishing a Scientific Research Paper*: Springer, Singapore.
  - Contains useful suggestions for writing the discussion and the conclusion sections of a scientific research paper.
Weissberg, R., and S. Buker. 1990. Writing up research: Prentice Hall Englewood Cliffs, NJ.
  - Discusses essential components of the discussion and the conclusion sections of a research paper and gives useful suggestions for them. Suggestions for writing other parts of a research paper are also given.

REFERENCES

Banerjee, S.S., M.V. Joshi, and R.V. Jayaram. 2006. Treatment of oil spills using organo-fly ash. *Desalination* 195 (1):32–39.
Boonseng, S., G.W. Roffe, M. Targema, J. Spencer, and H. Cox. 2017. Rationalization of the mechanism of in situ Pd(0) formation for cross-coupling reactions from novel unsymmetrical pincer palladacycles using DFT calculations. *J. Organomet. Chem.* 845:71–81.
Li, Y.-F., Y.-J. Sheng, and H.-K. Tsao. 2013. Evaporation stains: Suppressing the coffee-ring effect by contact angle hysteresis. *Langmuir* 29 (25):7802–7811.
Saha, S., B. Saint-Michel, V. Leynes, B.P. Binks, and V. Garbin. 2020. Stability of bubbles in wax-based oleofoams: Decoupling the effects of bulk oleogel rheology and interfacial rheology. *Rheol. Acta* 59 (4):255–266.
Ta, V.D., R.M. Carter, E. Esenturk, C. Connaughton, T.J. Wasley, J. Li, R.W. Kay, J. Stringer, P.J. Smith, and J.D. Shephard. 2016. Dynamically controlled deposition of colloidal nanoparticle suspension in evaporating drops using laser radiation. *Soft Matter* 12 (20):4530–4536.
Tyowua, A.T., O.O. Abel, S.O. Adejo, and M.E. Mbaawuaga. 2022. Functional properties of emulsified honey-vegetable oil mixtures. *ACS Food Sci. Technol.* 2 (3):581–591.
Tyowua, A.T., and B.P. Binks. 2021. Organic pigment particle-stabilized Pickering emulsions. *Colloids Surf. A Physicochem. Eng. Asp.* 613: 126044.

14 Summary of Research Paper Sections

Section	Main Function	Verb Tense	Contents
Title One-sentence summary of paper	Tells readers what the paper is all about as well as draws their attention	Present (recommended) or past tense	Keywords or phrases
Author(s) and affiliation(s) Names of significant contributors	Tells readers who did the work and where it was done		Names of people and names of their corresponding places of work
Textual abstract Multiple-sentence summary of paper (mini-version of paper)	Summarizes the paper (using sentences)	Present or past tense depending on the subsection	Brief background, aim of work, methods, important results and their ramifications and main conclusion(s)
Keywords Summary of paper in words and/or phrases	Summarizes the paper (using words and/or phrases)		Keywords and/or phrases (possibly those not in the title)
Graphical abstract An image summary of paper	Summarizes the paper (using an image or group of images)		Photographs, sketches, graphs, flowcharts and/or words

(Continued)

DOI: 10.1201/9781003186748-14

Section	Main Function	Verb Tense	Contents
Introduction Reason(s) for the research	Tells the reader the rationale for the study	Present tense (for established knowledge), present perfect tense (emphasize previous research in general terms) and simple past tense (emphasize specific findings)	Information that establishes the research area, statement that establishes a gap in the research area and statement that says how the research gap will be filled
Materials and Methods What was used and how it was used?	Gives a vivid description of the experimental procedure (*i.e.*, what was done)	Simple past tense (refers to what was done)	Detailed description of materials used and experimental procedures
Results What was obtained or found	Presents what was found (*i.e.*, data, either as figures or tables)	Simple past tense (refers to what was found)	Data obtained and observations during experiment
Discussion Meaning of what was obtained or found	Provides interpretation for what was found and puts it in the context of previous research	Past tense (to state a result) and present tense (to explain a result)	Interpretation of trends and relationship between data, comparison of data with previous research, generalizations and exceptions
Conclusion Take-home message	Summarizes the principle findings and gives the take-home message	Depends on the element being written	Answer to the aim or hypothesis, take-home message, supporting evidence for take-home message and ramifications of the take-home message
Acknowledgments List of helpers and funders	Says who helped with what	Simple past tense	Names of funders and those who helped with at various stages
References List of supporting or appropriate literature	Gives a list of relevant literatures consulted		List of all the relevant literature consulted during the research

FURTHER READING

Okuyama, Y. 2020. Use of Tense and Aspect in Academic Writing in Engineering: Simple Past and Present Perfect. *Journal of Pan-Pacific Association of Applied Linguistics* 24 (1):1–15.
- Contains useful suggestions regarding the use of simple past and present perfect tenses in scientific research papers.

15 Short Communication Papers

15.1 RESEARCH PAPERS *VERSUS* SHORT COMMUNICATION PAPERS

A short communication paper is a short and concise paper that reports an original and significant finding that needs to be disclosed urgently to a scientific community. A short communication paper can be thought of as an abridged version of a research paper because it has all the elements (*i.e.*, sections) of a research paper. However, these elements are not always clearly marked out in short communication papers, as illustrated in Example 15.1. These elements have the same functions, contents and language style as those of research papers. However, with short communication papers, their contents are concise and straight to the point. Some short communication papers marry some of these elements with others to keep their length short. This can be compared with research papers where these elements contain excruciating details and are clearly marked out. Also, while there are no strict word limits for research papers, short communication papers are subjected to a strict word limit of 2500–3500 depending on the journal.

Example 15.1 (p. 112–115)

Sample of a short communication paper. (From Vezza et al. 2021, *Chem. Commun.* 57(30), 3704–3707) The various sections are not clearly marked out, but they can be identified as shown. The experimental section is married with the results and the discussion sections.

15.2 WRITING A SHORT COMMUNICATION

Although your target journal's *Guide for Authors* is the best document to consult for advice on how to and how not to prepare your short communication paper, there are a few general helpful suggestions. Consider your short communication paper as an abridged research paper with the following elements: a title, author name(s) and address(es), abstracts (textual, graphical, video), an introduction, a materials and methods section, a results and discussion section, a conclusion section as well as an acknowledgment section and a list of references.

DOI: 10.1201/9781003186748-15

ChemComm

COMMUNICATION

View Article Online

View Journal | View Issue

🔖 Check for updates

Cite this: *Chem. Commun.*, 2021,
57, 3704

Received 18th February 2021,
Accepted 11th March 2021

DOI: 10.1039/d1cc00936b

rsc.li/chemcomm

An electrochemical SARS-CoV-2 biosensor inspired by glucose test strip manufacturing processes† | Title

Vincent J. Vezza, ⓘ *[a] Adrian Butterworth, ⓘ [a] Perrine Lasserre, ⓘ [a] Ewen O. Blair,[a] Alexander MacDonald, ⓘ [a] Stuart Hannah, ⓘ [a] Christopher Rinaldi,[b] Paul A. Hoskisson, ⓘ [c] Andrew C. Ward, ⓘ [d] Alistair Longmuir, ⓘ [e] Steven Setford,[e] Eoghan C. W. Farmer,[f] Michael E. Murphy[f,g] and Damion K. Corrigan ⓘ [a] | Authors

Abstract   Introduction Cont'd

Accurate and rapid diagnostic tests are critical to reducing the impact of SARS-CoV-2. This study presents early, but promising measurements of SARS-CoV-2 using the ACE2 enzyme as the recognition element to achieve clinically relevant detection. The test provides a scalable route to sensitive, specific, rapid and low cost mass testing.

SARS-CoV-2 came to the attention of health authorities during late 2019, shortly followed by declaration of a "public health emergency of international concern" on 30th of January 2020. SARS-CoV-2 quickly spread around the globe and was declared a pandemic by the World Health Organisation on 11th of March 2020. The virus, SARS-CoV-2, is the aetiological agent of coronavirus disease (Covid-19). Diagnostics have been a major priority and challenge thus far in the pandemic with assay quantity, reagent costs and time to result being of prime concern. SARS-CoV-2 has four major structural proteins,[1] with the spike protein known to bind to the surface of cells expressing angiotensin converting enzyme 2 (ACE2) on their surface. The affinity between ACE2 and the spike protein has been shown to be in the low nanomolar range[2] giving a similar level of affinity to an antibody-antigen interaction. The enzyme represents an important candidate molecule for construction of a biosensor because of the high affinity between the spike protein and ACE2

and the fact that a limited number of coronaviruses utilise ACE2 for entry (SARS-CoV-1, SARS-CoV-2 and HCoV-NL63) HCoV-NL63 causes infection primarily in young children, has a lower overall affinity for ACE2 (3–10 times lower)[3] and is a pathogen responsible for only mild/moderate childhood disease. It is therefore possible for ACE2 to be deployed as a selective receptor in various biosensor formats for this crucial category of human respiratory pathogen, either to allow definitive diagnosis of SARS-CoV-2 in adults or as a screening tool for identifying positive cases, which then receive confirmatory lab testing. Finally, the ACE2 enzyme is a carboxypeptidase responsible for the hydrolytic cleavage of angiotensin II to angiotensin I [1,7] liberating phenylalanine in the process. The active site is positioned away from SARS-CoV-2 binding site meaning that enzyme activity and binding from other proteins or small molecules in human samples is likely to be insignificant when compared to binding of an entire SARS-CoV-2 viral particle (50–200 nm) and the resulting effect on electrochemical signal. A critical feature of the ACE2 enzyme is the hydrophobic region which normally facilitates insertion into cell membranes[4] and allows ACE2 insertion into a synthetically made amphiphobic structures resembling cell membranes.

The fluorous effect is a well-known and well described tendency for fluorine atoms to avoid unfavoured interactions with other elements.[5] Formation of fluorous self-assembled monolayers (SAMs) have been utilised in organic electronics to reduce biological fouling of surfaces.[6] Here, a perfluorocarbon SAM formed of 1*H*,1*H*,2*H*,2*H*-Perfluorodecanethiol (PFDT) is deployed on a gold sensor surface to form a layer of amphiphobic character which facilitates straightforward insertion of ACE2 *via* its hydrophobic tail[7] (ordinarily employed for membrane insertion *in vivo*). Use of a perfluorocarbon SAM with its unique properties provided a robust combination of anti-biofouling behaviour and enzyme insertion, allowing straightforward sensor preparation.

Electrochemical biosensors offer a very attractive route to sensitive and low cost detection of biological analytes. The most

[a] *Biomedical Engineering, University of Strathclyde, Glasgow, G1 1XP, UK.*
  E-mail: vincent.vezza@strath.ac.uk
[b] *Pure & Applied Chemistry, University of Strathclyde, Glasgow, G1 1XP, UK*
[c] *Strathclyde Institute of Pharmacy and biomedical Sciences (SIPBS),*
  University of Strathclyde, Glasgow, G1 1XP, UK
[d] *Civil & Environmental Engineering, University of Strathclyde, Glasgow, G1 1XP,*
  UK
[e] *LifeScan Scotland Ltd, Beechwood Park North, Inverness, IV2 3ED, UK*
[f] *NHS GGC, Glasgow Royal Infirmary, Department of Microbiology,*
  NEW Lister Building, Glasgow, G31 2ER, UK
[g] *School of Medicine, Dentistry & Nursing, College of Medical Veterinary & Life*
  Sciences, University of Glasgow, G12 8QQ, UK
† Electronic supplementary information (ESI) available. See DOI: 10.1039/d1cc00936b

Communication   ChemComm

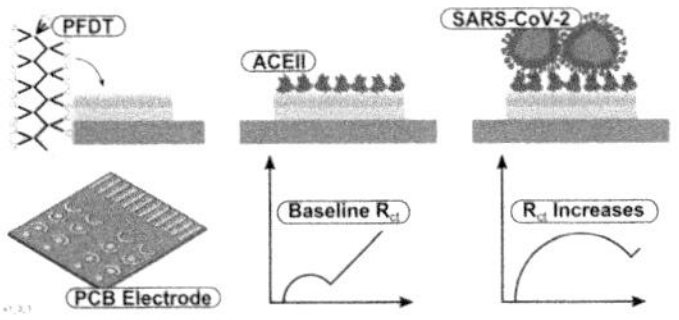

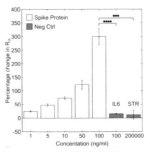

Scheme 1 Top – Two stage functionalisation process of the PFDT-ACE sensor and virus binding. Bottom – The electrode chip and the resulting R_{CT} increase upon virus recognition.

Introduction Cont'd

notable example of this is the glucose biosensor, operating *via* the glucose oxidase enzyme which is in widespread use each day by diabetic patients for routine determination of blood glucose levels. As a result of demand, facilities exist with very high volume manufacturing capability (*e.g.* several million sensors per day) and very well characterised chemical layer and enzyme deposition protocols. Due to the volume of sales in the glucose market (similar demand would exist for Covid-19 screening) unit costs of 20 UK pence[ii] can be achieved using established manufacturing processes. Very few diagnostic technologies, including lateral flow assays, can currently compete from a cost and volume perspective (*e.g.* current lateral flow Covid-19 tests cost £5–20 per test in the UK). In contrast, the sensor presented here requires a simple two stage preparation procedure of (1) PFDT deposition on a low cost PCB electrode and (2) ACE2 functionalisation through physisorbtion into the PFDT (Scheme 1). These steps are compatible with current manufacturing practices for glucose biosensors and importantly provide opportunity to rapidly translate the approach described here for SARS-CoV-2 detection into a glucose test strip production environment. This unlocks potential to achieve an extremely low cost Covid-19 assay, which utilises already CE marked substrates and approved assay readers, thus minimising the regulatory burden. Additionally, the assay is mutation proof as it exploits the interaction between SARS-CoV-2 spike protein and ACE2, has a degree of in-built surface orientation through the ability of ACE2 to insert *via* its hydrophobic region and offers the opportunity to develop similar tests for other respiratory viruses entering cells *via* membrane bound surface proteins.

PFDT-SAM and ACE2 functionalisation experiments were performed on a PCB sensor array bearing eight gold working electrodes. Such PCB arrays are low cost and allow high throughput assay development and represent an excellent test bed for this assay. When functionalised with PFDT and ACE2, the charge transfer resistance (R_{CT}) of the $[Fe(CN)_6]^{-3/-4}$ redox reaction, a signal parameter commonly used in impedimetric biosensor measurements, increased as expected (Fig. S1, ESI†). The PFDT SAM layer caused an increase in R_{CT} and a further increase was noted when ACE2 was physisorbed into the PFDT layer, confirming successful insertion of the enzyme. This was in contrast to experiments on facile ACE2 insertion with thiolated-hydrocarbon layers, where ACE2 insertion behaviour and sensor performance was not reproducible (Fig. S2, ESI†).

Fig. 1 Dose response curve for PFDT-ACE2 modified sensors when incubated for 30 min with solutions of recombinant SARS-CoV-2 spike protein. Negative controls: 100 ng mL^{-1} IL-6 and 200 µg mL^{-1} streptavidin. N = 8 & Error Bars = SE.

To characterise the sensitivity and specificity of the sensor, experiments to assess dose response behaviour and determine specificity of the ACE2 spike protein interaction were conducted using recombinant spike protein, streptavidin and IL-6. IL-6, which is found at elevated levels in Covid-19 patients, and streptavidin, widely used in bioassays, were employed as negative controls. A clear dose response effect was found for recombinant spike-protein binding to ACE2 (Fig. 1). Fitting the dose response curve gave a limit of detection for recombinant spike protein of 1.68 ng mL^{-1}. The binding between ACE2 and spike protein at 100 ng mL^{-1} was shown to be significantly different to IL-6 (100 ng mL^{-1}, P = 3.158E-7) and streptavidin (200 µg mL^{-1}, P = 3.981E-4) binding. This suggests it was possible to construct a sensitive and selective sensor operating on the principle of ACE2-spike glycoprotein interactions. Note – example Nyquist plots, equivalent circuit and representative values are shown in Fig. S3 (ESI†).

Next, assay performance in more complex samples was explored. Due to biosafety restrictions, the sensor was tested with inactivated virus samples at different concentrations alongside a negative sample from a commercial kit. All samples were provided in viral transport medium which contains proteins and dead eukaryotic cells used to culture the virus and was an appropriate proxy for a complex human tissue such as saliva. Furthermore, these kits are used to validate diagnostic tests in clinical biomedical laboratories because of their close resemblance to clinical samples.[9] When a viral dose response curve was constructed, it was possible to detect the inactivated virus with increasing concentration (Fig. 2). Curve fitting and comparison to blank solution yielded an LoD of 38.6 copies mL^{-1}. At 10^4 and 10^5 copies mL^{-1}, the impedance change was significantly different to the negative control (P = 0.004 and P = 6.765E-4 respectively). The background signal increase observed in the negative control is most likely due to

Experimental with Results & Discussion

ChemComm Experimental with Results & Discussion Communication

View Article Online

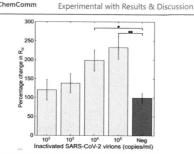

Fig. 2 Dose response following incubation of PFDT-ACE2 modified sensor with inactivated SARS-CoV-2 molecular standards kit. Viral concentration is expressed in copies mL^{-1} which is the result of sample quantification following viral inactivation using digital PCR. N = 6 & Error Bars = SE.

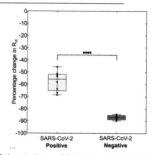

Fig. 3 Assay signal response following incubation with one SARS-CoV-2 positive patient sample (Ct = 26) and a negative sample. N = 8, horizontal line = mean.

residual non-specific surface interactions from the complex viral sample medium (proteins, cellular debris *etc.*). To provide context, RNA levels for SARS-CoV-2 in saliva samples have been estimated in the range of 10^4–10^{11} copies per mL.[10,11] Therefore, these findings show it was possible to detect inactivated SARS-CoV-2 from a complex sample in 30 minutes at clinically relevant detection levels. Coupling this to the sensor being designed for manufacture in mass production environments, it raises the possibility of developing a Covid-19 assay using established sensors and instrumentation from the blood glucose industry.

To assess performance of the sensor more comprehensively, two human saliva samples were tested in bio secure conditions. One SARS-CoV-2 positive human saliva sample, confirmed through PCR based clinical diagnostics (Ct = 26 cycles, indicating high viral load), was transferred into inactivation solution called VPSS containing a proprietary cell culture medium and Triton-X (to denature the viral envelope whilst leaving proteins largely intact). This was compared against a negative human saliva sample (confirmed *via* PCR) which had undergone the same handling processes as the positive sample. To carry out this experiment an improved assay protocol was used. Firstly, the positive and negative samples were premixed with ACE2 and then allowed to physisorb onto the PFDT modified sensor surface with a measurement taken immediately after sample addition. This increased the practicality of the assay by accelerating time to result and has the potential to make storage, transport easier and shelf life longer because sensor strips can be shipped without the ACE2 enzyme being preprinted. Secondly, the clinical samples were tested in VPSS medium. Once inactivated in VPSS, suspected SARS-CoV-2 positive samples can be manipulated at Biosafety Level 2 in contrast to untreated patient samples, or samples in Viral Transport Medium (VTM), which require Biosafety Level 3 facilities. The ability to detect SARS-CoV-2 in VPSS using the biosensor reported here means that as well as for self testing,

the assay is compatible with more centralised testing facilities where it is important to protect an operator running multiple samples such as hospital wards, GP surgeries, care homes, airports and workplaces. Adopting this revised approach, the R_{CT} changes in the positive and negative samples were found to be significantly different (Fig. 3, $P = 1.2 \times 10^{-7}$). Changes to the assay approach with clinical samples resulted in an apparent drop in R_{CT}. This is attributed to a baseline artefact caused by Triton-X and sensor surface interactions in VPSS. Further investigation is being performed to fully understand and quantify these effects.

Numerous SARS-CoV-2 biosensor approaches have been developed and reported recently. Magnetic beads, carbon black electrodes and multiple antibodies were employed to achieve a detection limit of 6.5 plaque forming units mL^{-1} in whole virus.[12] An ultrasensitive super sandwich-type assay has been presented to detect viral RNA from SARS-CoV-2 with an LoD of 200 copies mL^{-1},[13] and a FET biosensor capable of detecting spike protein with a LoD of 242 copies mL^{-1} in transport medium has also been reported.[9] Plasmonic photothermal biosensors have been shown to detect down to 0.22 pM of viral RNA, which was estimated to be around 2.26×10^4 copies.[14] Other electrochemical sensors have also shown good sensitivity, including molecular imprinted polymers (15 fM)[15] and graphene and gold nanostars (1.68×10^{-22} µg mL^{-1}).[16] The sensor developed here has comparable and clinically relevant sensitivity (LoD for inactivated virus of 38.6 copies mL^{-1}), with a clear route to manufacture. Importantly, the design of the sensor reported here is resistant to mutations in the spike protein. In terms of the chosen surface chemistry and the use of PFDT in a SAM layer, there are examples in the literature of other sensor systems with good anti-biofouling properties and these have been well reviewed.[17] Demonstrations of anti-fouling properties in ternary SAMs have recently been made[18,19] and there are advantages to these approaches for more complex sensors. However, in our experiments we could not reproduce the ACE2 insertion behaviour

View Article Online

with standard hydrocarbon SAMs (Fig. S2, ESI†) and because of the emphasis on eventual sensor manufacture, PFDT was found to be well suited to the end application where monolayer coverage was sufficient to both retain the enzyme *via* amphiphobic interactions and provide anti-fouling properties.

Specificity of this sensor has been demonstrated against IL6, streptavidin, virial transport medium and one negative clinical saliva sample. This suggests that biofouling and non-specific binding of other commonly encountered biological proteins does not compromise assay sensitivity or specificity. Another consideration is other viral pathogens that bind to ACE2. Of the three viruses that are known to bind to ACE2, SARS-CoV-2 has the greatest affinity. SARS-CoV-1 is not present in the community and is therefore unlikely to impact the performance of this sensor. HCoV-NL63 primarily affects paediatric patient groups, which could limit the sensor to adult diagnostics if signal stratification is not possible. The current time to result based on the presented results is 30 min, which is competitive with existing SARS-CoV-2 assays but significant scope exists to reduce our assay time to faster sample turnaround times.

Further work will include sensor verification with a large set of clinical samples to determine, (1) whether low, medium and high viral load SASR-CoV-2 samples can be stratified alongside HCoV-NL63 samples or whether positive vs negative is the only feasible assay output, (2) fully understanding the measurement artefact effects of surfactants and detergents from VPSS and VTM, (3) testing shorter incubation times to reduce the overall assay time to result below the current 30 minutes and (4) testing simpler electrochemical measurements (*e.g.* amperometry) to determine if more straight forward measurement circuits (and therefore a lower cost reader can be employed) than those required for impedance.

These results present the basis of a novel, sensitive, selective and scalable biosensor assay which is based upon the principle of immobilising ACE2 into a layer with amphiphobic character. This sensor assay has sensitivity (1.68 ng mL^{-1}) with recombinant spike protein and specificity against two proteins (IL-6 and streptavidin). When evaluated with a SARS-CoV-2 analytical grade laboratory inactivated virus kit, an LoD of 38.6 copies mL^{-1} was achieved. Finally, testing with high viral load clinical sample (Ct = 26) showed strong discrimination against a negative sample following inactivation in VPSS medium. By taking advantage of glucose biosensor production approaches (per sensor cost > £0.2) it is possible to demonstrate a clear path to translation through established manufacturing techniques, utilisation of existing sensor substrates and read out devices with prior CE marking. These findings together point towards a useful diagnostic tool, which can be rapidly deployed at low cost to screen for SARS-CoV-2 in a range of settings.

Conflicts of interest

There are no conflicts to declare.

Notes and references

1 P. S. Masters, *Adv. Virus Res.*, 2006, **65**, 193–292.
2 J. Lan, J. Ge, J. Yu, S. Shan, H. Zhou, S. Fan, Q. Zhang, X. Shi, Q. Wang, L. Zhang and X. Wang, *Nature*, 2020, **581**, 215–220.
3 A. C. Mathewson, A. Bishop, Y. Yao, F. Kemp, J. Ren, H. Chen, X. Xu, B. Berkhout, L. van der Hoek and I. M. Jones, *J. Gen. Virol.*, 2008, **89**, 2741–2745.
4 F. J. Warner, A. I. Smith, N. M. Hooper and A. J. Turner, *Cell. Mol. Life Sci.*, 2004, **61**, 2704–2713.
5 M. Cametti, B. Crousse, P. Metrangolo, R. Milani and G. Resnati, *Chem. Soc. Rev.*, 2012, **41**, 31–42.
6 Q. J. Cai, M. B. Chan-Park, Q. Zhou, Z. S. Lu, C. M. Li and B. S. Ong, *Org. Electron.*, 2008, **9**, 936–943.
7 P. Towler, B. Staker, S. G. Prasad, S. Menon, J. Tang, T. Parsons, D. Ryan, M. Fisher, D. Williams, N. A. Dales, M. A. Patane and M. W. Pantoliano, *J. Biol. Chem.*, 2004, **279**, 17996–18007.
8 National Institute for Health and Care Excellence, Blood glucose testing strips, https://bnf.nice.org.uk/medical-device-type/blood-glucose-testing-strips-2.html, 2021.
9 G. Seo, G. Lee, M. J. Kim, S.-H. Baek, M. Choi, K. B. Ku, C.-S. Lee, S. Jun, D. Park, H. G. Kim, S.-J. Kim, J.-O. Lee, B. T. Kim, E. C. Park and S. I. Kim, Correction to Rapid Detection of COVID-19 Causative Virus (SARS-CoV-2) in Human Nasopharyngeal Swab Specimens Using Field-Effect Transistor-Based Biosensor, *ACS Nano*, 2020, DOI: 10.1021/acsnano.0c02823.
10 Y. M. Bar-On, A. Flamholz, R. Phillips and R. Milo, SARS-CoV-2 (COVID-19) by the numbers, *eLife*, 2020, DOI: 10.7554/eLife.57309.
11 R. Wölfel, V. M. Corman, W. Guggemos, M. Seilmaier, S. Zange, M. A. Müller, D. Niemeyer, T. C. Jones, P. Vollmar, C. Rothe, M. Hoelscher, T. Bleicker, S. Brünink, J. Schneider, R. Ehmann, K. Zwirglmaier, C. Drosten and C. Wendtner, *Nature*, 2020, **581**, 465–469.
12 L. Fabiani, M. Saroglia, G. Galatà, R. De Santis, S. Fillo, V. Luca, G. Faggioni, N. D'Amore, E. Regalbuto, P. Salvatori, G. Terova, D. Moscone, F. Liana and F. Arduini, *Biosens. Bioelectron.*, 2021, **171**, 112686.
13 H. Zhao, F. Liu, W. Xie, T. C. Zhou, J. OuYang, L. Jin, H. Li, C. Y. Zhao, L. Zhang, J. Wei, Y. P. Zhang and C. P. Li, *Sens. Actuators, B*, 2021, **327**, 128899.
14 G. Qiu, Z. Gai, Y. Tao, J. Schmitt, G. A. Kullak-Ublick and J. Wang, *ACS Nano*, 2020, **14**, 5268–5277.
15 S. A. Hashemi, N. G. Golab Behbahan, S. Bahrani, S. M. Mousavi, A. Gholami, S. Ramakrishna, M. Firoozsani, M. Moghadami, K. B. Lankarani and N. Omidifar, *Biosens. Bioelectron.*, 2021, **171**, 112731.
16 A. Raziq, A. Kidakova, R. Boroznjak, J. Reut, A. Öpik and V. Syritski, *Biosens. Bioelectron.*, 2021, **178**, 113029.
17 A. Hassan and L. M. Pandey, *Polym.-Plast. Technol. Eng.*, 2015, **54**, 1358–1378.
18 A. Miodek, E. M. Regan, N. Bhalla, N. A. E. Hopkins, S. A. Goodchild and P. Estrela, *Sensors*, 2015, **15**, 25015–25032.
19 S. Campuzano, F. Kuralay, M. J. Lobo-Castañón, M. Bartošík, K. Vyavahare, E. Paleček, D. A. Haake and J. Wang, *Biosens. Bioelectron.*, 2011, **26**, 3577–3583.

Experimental with Results & Discussion

Conclusions

ChemComm

COMMUNICATION

View Article Online
View Journal

Check for updates

Cite this: DOI: 10.1039/d2cc02290g

Received 23rd April 2022,
Accepted 7th June 2022

DOI: 10.1039/d2cc02290g

rsc.li/chemcomm

An iodine-containing probe as a tool for molecular detection in secondary ion mass spectrometry†

Selda Kabatas Glowacki, ‡[ab] Paola Agüi-Gonzalez, ‡[ab]
Shama Sograte-Idrissi [ab] Sebastian Jähne, [b] Felipe Opazo [a]
Nhu T. N. Phan §[ab] and Silvio O. Rizzoli *[ab]

We developed here an iodine-containing probe that can be used to identify the molecules of interest in secondary ion mass spectrometry (SIMS) by simple immunolabelling procedures. The immunolabelled iodine probe was readily combined with previously-developed SIMS probes carrying fluorine, to generate dual-channel SIMS data. This probe should provide a useful complement to the currently available SIMS probes, thus expanding the scope of this technology.

Mass spectrometry imaging (MSI) was introduced to biology a few decades ago, as an imaging technique able to reveal the chemical composition of tissues and cells, and has improved steadily, reaching very high levels of precision.[1-3] To analyse the chemical species on a sample and estimate their abundance, the primary ion beam of a secondary ion mass spectrometer is focused on the surface of the sample, causing the ejection of secondary particles by the impact of the primary ions.[4] A broad spectrum of such instruments is available, with the highest lateral resolutions being provided by the NanoSIMS implementation, at 50 nm or lower.[5] However, the intrinsic nature of such an instrument implies that only mono- or diatomic molecules are detected. Further chemical information is lost, due to the extensive fragmentation induced by the highly reactive primary ion sources (Cs^+/O^-).[6] This problem also affects other SIMS implementations, such as time-of-flight-SIMS (TOF-SIMS), which can reveal molecules only up to ~2.5 kDa.[3]

Thus, labelling particular molecules using SIMS-compatible probes is crucial to detect and visualize larger species, such as proteins. To make this possible, the probes must contain stable elements that are easily ionized, and thus provide a good secondary ion yield. A few different probes fulfilling these requirements have been successfully developed in the last few years. The first investigations employed heavy metal particles such as gold[7] or lanthanides.[8] Further efforts were then focused on reducing the size of the probes, to go along with the continuous improvements in the spatial resolution of SIMS. Click-chemistry tags containing fluorine or ^{15}N,[9,10] as well as gold nanoparticles- or fluorine-conjugated nanobodies[11,12] have also been successfully applied. Boron-containing probes, developed for the positive secondary ion mode of SIMS, have been recently produced and applied.[13]

However, few multi-targeted proteins SIMS imaging experiments have been performed in the last decades. Due to their low ionization energy,[14] lanthanides are suitable for their detection in the positive ion mode, and enable multi-protein analysis,[8] albeit the large size of their particles limits labelling efficiency. Fluorine or gold are better candidates for the negative ion mode and could be used, in principle, to simultaneously reveal the localization of two proteins of interest with SIMS, but such experiments have not yet been performed. Iodine is the element with the fourth highest electron affinity, after fluorine, chlorine and bromine.[15] Thus, this element represents an excellent choice as SIMS reporter for the negative ion mode and would enable a straightforward strategy for dual-targeted proteins SIMS imaging in the negative secondary ion mode, when combined with another compatible probe. We therefore decided here to develop iodine-conjugated probes for SIMS imaging.

Initially we focused on a precursor molecule with a structure similar to the pentafluorobenzoyl groups in FluorLink, which we developed for NanoSIMS use in the past.[11] However, approaches with pentaiodo-benzoic acid or with reduced iodine content using tetraiodophthalic anhydride and triiodobenzoic acid did not lead to the desired products when coupled to a short peptide at its N-terminus. We had determined iodine loss by ESI-HRMS, which can be explained simply by the lower bond dissociation energy of iodine–carbon bonds and their

[a] Center for Biostructural Imaging of Neurodegeneration, University Medical Center Göttingen, Von-Siebold-Straße 3a, 37075 Göttingen, Germany.
E-mail: srizzol@gwdg.de
[b] Department of Neuro and Sensory Physiology, University Medical Center, Göttingen, Humboldtalee 23, 37073 Göttingen, Germany
† Electronic supplementary information (ESI) available. See DOI: https://doi.org/10.1039/d2cc02290g
‡ Both authors contributed equally to this manuscript.
§ Current address: Department of Chemistry and Molecular Biology, University of Gothenburg, Kemivägen 10, SE-412 96 Gothenburg, Sweden.

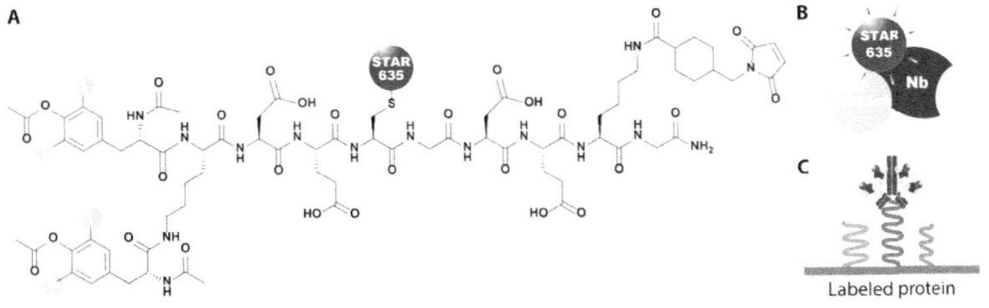

View Article Online

Communication

ChemComm

Fig. 1 Schematic representation of immunolabeling of proteins of interest by using IodLink-nanobody anti-mouse κIg. (A) IodLink consists of a soluble peptide with diiodo-L-thyroxine groups on the N-terminus, a fluorophore on the cysteinyl side chain and a maleimide group on the lysine side chain. (B) The thiol-reactive IodLink is conjugated to a secondary nanobody that detects a primary antibody binding the protein of interest. (C) A sequence of primary, secondary and even tertiary antibodies can be used to enhance the signal, increasing the number of IodLink nanobodies in the respective area.

instability under non-inert, oxidative conditions. We therefore searched for iodinated compounds in natural systems and chose diiodo-L-tyrosine, the precursor compound for the synthesis of L-thyroxine, a thyroid hormone essential for vertebrates.[16]

The synthesis was similar to previously published methods,[9–11] beginning with amino acid coupling on a solid support (ESI,† Section S3). The peptide was synthesized from the C- to the N-terminus on an acid-sensitive Sieber amide resin using standard fluorenylmethyloxycarbonyl solid-phase peptide synthesis (Fmoc-SPPS). After coupling of the last amino acid, Fmoc-Lys(Fmoc)-OH, the Fmoc groups were deprotected by basic treatment. Then the commercially available N,O-diacetyl-3,5-diiodo-L-tyrosine was linked to the two amino groups of the N-terminal lysine. The final peptide was cleaved from the solid support under acidic conditions. In the next step, the fluorophore Star635 maleimide was coupled to the cysteine side chain by a sulfhydryl-maleimide reaction. Finally, a maleimide linker was added to the lysine side chain to generate IodLink with $4\times^{127}I$ atoms per molecule (Fig. 1A). Each reaction step at solid support was analyzed by HPLC and verified by HR-ESI-MS with no iodine loss observed (ESI,† Section S7). The stability of the iodinated peptide was also confirmed in MALDI-ToF measurements applying different laser powers (ESI,† Section S7). After identification, the products of each step were purified by HPLC to achieve purities of $\geq 95\%$. IodLink was then conjugated to a secondary nanobody directed against the kappa (κ) light chain of mouse immunoglobulins (κIg), enabling it to be used in immunocytochemistry experiments (ESI,† Section S4) (Fig. 1B). The selected nanobody has two ectopic cysteines that react with IodLink molecules in a sulfydryl-maleimide reaction, resulting in a doubling of the amount of ^{127}I per detected target of interest.

The probe is thereby designed in a fashion that enables it to be detected both in SIMS and in fluorescence microscopy, to enable users to optimize their sample labelling using fluorescence imaging, before turning to the more laborious SIMS experiments.

After accomplishing the nanoprobe conjugation, we performed tests to optimize the labelling with IodLink-nanobody

anti-mouse κIg. For this, we applied immunostainings on COS-7 cells with a mouse anti-α-tubulin as primary antibody and varying concentrations of IodLink-nanobody (ESI,† Fig. S2). We found that 20 nM IodLink-nanobody provides an adequate signal while maintaining a lower background level.

After demonstrating the nanoprobe's detectability and specificity by fluorescence microscopy, we turned to immunostaining on COS-7 cells, repeating the same approach as for α-tubulin but targeting the peroxisomal PMP70 protein, which is known to provide a very clear and easily recognized pattern in SIMS microscopy.[11] To increase the signal intensity to the maximum possible, we employed a signal enhancement procedure in which we immunostained the cells with a rabbit polyclonal antibody for PMP70, followed by a secondary goat anti-rabbit antibody, and finally a mouse anti-goat antibody, to which the IodLink-nanobody anti-mouse κIg was bound (Fig. 1C). The samples were then embedded in resin, sliced to a thickness of 200 nm and imaged in NanoSIMS. To obtain negative secondary ions, we selected $^{133}Cs^+$ as primary ion source and set several detectors to collect $^{12}C^{14}N^-$, $^{32}S^-$, and $^{127}I^-$ respectively (Fig. 2). $^{12}C^{14}N^-$ is the most common ion employed to visualize biological samples with NanoSIMS, providing an accurate image of nitrogen-rich cellular material such as proteins or nucleic acids.[17]

We also included $^{32}S^-$ to obtain additional information and facilitate the location of the cellular nuclei, due to its high abundance in structures such as the heterochromatin or nucleoli.[18,19] Finally, we collected $^{127}I^-$ ion signal to locate and visualize our proteins of interest labelled with IodLink-nanobody anti-mouse κIg (Fig. 2).

The iodine signal reproduces the typical distribution of the peroxisomes, showing a significantly higher signal in small dotted regions that are exclusively found in the cytoplasm, never penetrating the nucleus (Fig. 2). The unspecific signal obtained in negative control samples was negligible (Fig. 2 and ESI,† Fig. S3), suggesting that the IodLink-nanobody provides high specific labeling and low background signal for NanoSIMS imaging.

To further test the possibilities of this new nanoprobe, we performed an immunostaining for two proteins of interest, by

This journal is © The Royal Society of Chemistry 2022

A Practical Guide to Scientific Writing in Chemistry

View Article Online

ChemComm

Communication

Fig. 2 NanoSIMS images of the peroxisomal PMP70 immunostained with IodLink-nanobody anti-mouse κIg. On the upper row, a cell immunostained to detect PMP70 is indicated. On the lower row, a negative control cell is shown. The negative control includes the primary antibody and IodLink-nanobody anti-mouse κIg, but skipping the incubations with the intermediate antibodies. From left to right: $^{12}C^{14}N^-$, $^{32}S^-$, $^{127}I^-$ images, overlay of $^{12}C^{14}N^-$ (grey) and $^{127}I^-$ (green), and HSI image of $^{127}I^-/^{12}C^{14}N^-$ (colour scale: magenta = maximum; blue = minimum). Scale bars: 5 μm.

combining IodLink-nanobody anti-mouse κIg with the FluorLink-nanobodies that we developed in the past.[11] The mass dispersion range offered by the NanoSIMS 50L instrument allows the simultaneous detection of $^{127}I^-$ and other mass peaks, such as $^{12}C^{14}N^-$ and $^{19}F^-$, which enabled us to analyse the mitochondrial marker (TOM70) linked to a green fluorescent protein (GFP) on COS-7 cells, using FluorLink-nanobodies anti-GFP #1 and #2, while we employed IodLink-nanobody to label the peroxisomal marker PMP70 (Fig. 3). Both probes could be introduced simultaneously through a single immunolabelling incubation step. The resulting negative ion mode NanoSIMS images showed that this procedure determines with relative ease the distribution of two proteins of interest. The regions enriched with ^{127}I and ^{19}F indicated that there were no interferences between FluorLink and IodLink probes as they showed typical patterns for mitochondrial and peroxisomal markers, respectively (Fig. 3).

The probes seem to be highly specific, but one should still take into account the fact that IodLink contains 3 fluorine

atoms delivered from the fluorophore Star635. The doubling of IodLink after coupling to the nanobody's two ectopic cysteines results in 6 fluorine atoms, which in principle could interfere with our capacity to distinguish the location of two proteins of interest if combined with FluorLink-nanobodies which deliver 52xF per target. To check this, we tested the correlation level between $^{127}I^-$ and $^{19}F^-$ within the cells, using the correlation between ^{32}S and $^{12}C^{14}N$ as an internal reference (ESI,† Fig. S4). With this analysis, we observed the expected a strong correlation between $^{32}S^-$ and $^{12}C^{14}N^-$, and a very weak, not significant correlation between $^{127}I^-$ and $^{19}F^-$. These results demonstrate that the signal obtained from IodLink is clearly distinguishable from the FluorLink. This is in line with our previous observations, when we noticed that NanoSIMS imaging cannot solely rely on the fluorine atoms of the fluorophore Star635 to reveal the localization of proteins of interest, due to its low number of F atoms.

These experiments imply that iodine-based SIMS imaging is feasible, and can also be employed in multi-targeted proteins

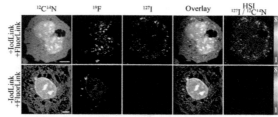

Fig. 3 NanoSIMS images of a dual-isotope immunostaining using FluorLink- and IodLink-nanobodies. TOM70-GFP transfected COS-7 cells were labelled with FluorLink-nanobodies against GFP, and the protein PMP70 was immunostained with antibodies that were detected using the IodLink-nanobody anti-mouse κIg, as above. From left to right: $^{12}C^{14}N^-$, $^{19}F^-$, $^{127}I^-$ images, overlay of $^{12}C^{14}N^-$ (grey), $^{19}F^-$ (magenta) and $^{127}I^-$ (green), and HSI image of $^{127}I^-/^{12}C^{14}N^-$ (colour scale: magenta = maximum; blue = minimum). On the upper row, representative cells stained with the entire sequence of antibodies. The lower row represents the negative control, which was labelled with the fluorine probe, as well as with the primary antibody and IodLink-nanobody anti-mouse κIg, but excludes the intermediate antibodies for the IodLink immunostaining. Scale bars: 5 μm.

Chem. Commun.

imaging with NanoSIMS by combining it with other elements, detectable in negative ion mode, such as fluorine. In comparison to other elements, iodine offers a very sparse endogenous presence within biological samples. Fluorine, for example, is a prominent element in the production of detergents and plastics, including the common tubes and reagent holders used in biomedical sciences.[20] This implies that the background level for iodine tends to be much lower than for fluorine, as we can appreciate in the double staining images, in which a low fluorine signal can be observed throughout the samples.

This work demonstrates the synthesis of IodLink, a non-radioactive, stable iodine-containing probe, which can be conjugated to different surfaces or proteins bearing an accessible thiol group. To show its applicability, we successfully conjugated IodLink to an anti-mouse-nanobody, indicating the feasibility of labelling specific proteins of interest, as well as the visualization and location of such proteins by both fluorescence microscopy and SIMS.

In our work, we also demonstrated that IodLink-nanobody was effective for specific protein imaging with NanoSIMS at subcellular resolution due to its high specificity and high signal-to-noise ratio, even in dual-protein targeted imaging. Future experiments combining additional probes, also detectable on the negative secondary ion mode of SIMS, such as gold-conjugated nanobodies, could further increase the number of proteins simultaneously visualized by SIMS. In the same manner, all of these probes are compatible with the use of other isotopes such as ^{13}C, ^{15}N or ^{18}O, to study in parallel different cellular metabolic processes, such as the turnover of proteins or lipids.[21–24] Importantly, IodLink could also be detected by other SIMS instruments such as ToF-SIMS. Likewise, IodLink could be tested on other imaging techniques such as X-ray scattering, implying that this probe, while only showcased here for its application to NanoSIMS, should enable substantially more applications in the future.

S. K. G. and S. O. R. conceptualized the project. S. K. G. performed all chemistry work. S. S. I. performed the immuno-labelling experiments and fluorescence imaging. F. O provided materials and supervised nanobody conjugation and their applications in IF. S. J. performed the plastic embedding. P. A. G. performed the SIMS imaging and N. T. N. P. supervised all SIMS work. P. A. G. and S. O. R. analysed the data. S. K. G., P. A. G. and S. O. R. wrote the initial draft.

We thank Prof. Dr Ulf Diederichsen (Institute of Organic and Biomolecular Chemistry, University of Göttingen, Germany) for the generous support by using his infrastructure. We also acknowledge the service department of the faculty of chemistry (University Göttingen, Germany), especially Dr. Holm Frauendorf and his team for the mass spectrometric measurements. This work was supported by grants to S. O. R. from the German Research Foundation (Deutsche Forschungsgemeinschaft, DFG SFB1286/A03, RI 1967/7-3, RI 1967/11-1) and from the Nieders. Vorab (76251-12-6/19/ZN 3458). Also supported under Germany's Excellence Strategy-EXC 2067/1-390729940. Silvio O. Rizzoli and Felipe Opazo have received compensation as consultants of NanoTag Biotechnologies GmbH and own stocks in the company.

Conflicts of interest

The other authors declare no competing interests.

Notes and references

1 M. J. Taylor, J. K. Lukowski and C. R. Anderton, *J. Am. Soc. Mass Spectrom.*, 2021, **32**, 872–894.
2 A. P. Bowman, R. M. A. Heeren and S. R. Ellis, *TrAC, Trends Anal. Chem.*, 2019, **120**, 115197.
3 P. Agüi-Gonzalez, S. Jähne and N. T. N. Phan, *J. Anal. At. Spectrom.*, 2019, **34**, 1355–1368.
4 S. G. Boxer, M. L. Kraft and P. K. Weber, *Annu. Rev. Biophys.*, 2009, **38**, 53–74.
5 J. Malherbe, F. Penen, M.-P. Isaure, J. Frank, G. Hause, D. Dobritzsch, E. Gontier, F. Horréard, F. Hillion and D. Schaumlöffel, *Anal. Chem.*, 2016, **88**, 7130–7136.
6 B. L. Gorman and M. L. Kraft, *Anal. Chem.*, 2020, **92**, 1645–1652.
7 R. L. Wilson, J. F. Frisz, W. P. Hanafin, K. J. Carpenter, I. D. Hutcheon, P. K. Weber and M. L. Kraft, *Bioconjug. Chem.*, 2012, **23**, 450–460.
8 M. Angelo, S. C. Bendall, R. Finck, M. B. Hale, C. Hitzman, A. D. Borowsky, R. M. Levenson, J. B. Lowe, S. D. Liu, S. Zhao, Y. Natkunam and G. P. Nolan, *Nat. Med.*, 2014, **20**, 436–442.
9 I. C. Vreja, S. Kabatas, S. K. Saka, K. Kröhnert, C. Höschen, F. Opazo, U. Diederichsen and S. O. Rizzoli, *Angew. Chem., Int. Ed.*, 2015, **54**, 5784–5788.
10 S. Kabatas, I. C. Vreja, S. K. Saka, C. Höschen, K. Kröhnert, F. Opazo, S. O. Rizzoli and U. Diederichsen, *Chem. Commun.*, 2015, **51**, 13221–13224.
11 S. Kabatas, P. Agüi-Gonzalez, R. Hinrichs, S. Jähne, F. Opazo, U. Diederichsen, S. O. Rizzoli and N. T. N. Phan, *J. Anal. At. Spectrom.*, 2019, **34**, 1083–1087.
12 P. Agüi-Gonzalez, T. M. Dankovich, S. O. Rizzoli and N. T. N. Phan, *Nanomaterials*, 2021, **11**, 1797.
13 S. Kabatas, P. Agüi-Gonzalez, K. Saal, S. Jähne, F. Opazo, S. O. Rizzoli and N. T. N. Phan, *Angew. Chem., Int. Ed.*, 2019, **58**, 3438–3443.
14 P. F. Lang and B. C. Smith, *J. Chem. Educ.*, 2010, **87**, 875–881.
15 T. Andersen, H. K. Haugen and H. Hotop, *J. Phys. Chem. Ref. Data*, 1999, **28**, 1511–1533.
16 C. E. Citterio, H. M. Targovnik and P. Arvan, *Nat. Rev. Endocrinol.*, 2019, **15**, 323–338.
17 M. L. Steinhauser and C. P. Lechene, *Semin. Cell Dev. Biol.*, 2013, **24**, 661–667.
18 A. A. Legin, A. Schintlmeister, M. A. Jakupec, M. Galanski, I. Lichtscheidl, M. Wagner and B. K. Keppler, *Chem. Sci.*, 2014, **5**, 3135–3143.
19 F. Lange, P. Agüi-Gonzalez, D. Riedel, N. T. N. Phan, S. Jakobs and S. O. Rizzoli, *PLoS One*, 2021, **16**, e0240768.
20 S. Banerjee, *Handbook of Specialty Fluorinated Polymers*, Elsevier, 2015.
21 S. K. Saka, A. Vogts, K. Kröhnert, F. Hillion, S. O. Rizzoli and J. T. Wessels, *Nat. Commun.*, 2014, **5**, 3664.
22 S. Jähne, F. Mikulasch, H. G. H. Heuer, S. Truckenbrodt, P. Agüi-Gonzalez, K. Grewe, A. Vogts, S. O. Rizzoli and V. Priesemann, *Cell Rep.*, 2021, **34**, 108841.
23 M. L. Kraft and H. A. Klitzing, *Biochim. Biophys. Acta*, 2014, **1841**, 1108–1119.
24 C. He, X. Hu, R. S. Jung, T. A. Weston, N. P. Sandoval, P. Tontonoz, M. R. Kilburn, L. G. Fong, S. G. Young and H. Jiang, *Proc. Natl. Acad. Sci. U. S. A.*, 2017, **114**, 2000–2005.

a. *Title*
- Avoid question titles as much as possible.
- The title should be short ($\leq$ 15 words), simple and informative with the content clearly reflected.
- Use keywords and phrases as much as possible to facilitate discoverability of the paper in online searches.

b. *Author name(s) and address(es)*
- List the names of all the authors as well as their corresponding addresses.
- Designate one author for correspondence.

c. *Graphical abstract*
- This may consist of a small graphic and a brief text (15–25 words) depending on the journal.
- Use an informative eye-catching graphic with no more than two elements (see Chapter 7).
- The text should state the main findings and their importance. Use words and phrases that are easily understandable.

d. *Textual and video abstracts*
- The textual abstract should contain keywords and phrases, and it should be short ($\leq$ 100 words), simple and easily understandable.
- State the specific objective or aim, the main findings as well as their ramifications.
- Prepare video abstracts using the same guide given for video abstracts of scientific research papers in Chapter 5.

e. *Introduction*
- The introduction should be brief and concise.
- It should contain a brief background, a statement of the problem and the specific objective or aim.
- Avoid a detailed literature review.

f. *Materials and methods*
- Should be brief and concise, with sufficient information to enable independent reproduction of the results.

g. *Results and discussion*
- Should contain key results.
- Interpret the results and discuss their significance.

h. *Conclusion*
- Should contain the take-home message.
- Give a brief summary of the main results and state their implications.

i. *Acknowledgments*
- Acknowledge the contributions of individuals and organizations toward the success of the work, including those that funded it.

15.3   REFERENCES

Unlike research papers where there is no strict limit to the number of references, short communications papers are given a limited number of references, usually fifteen or so, depending on the journal.

A sample of a short communication paper is shown in Example 15.1. The various sections discussed here are not clearly marked out in the example, but they can be identified as shown. More samples of short communication papers can be found in "Chemical Communications", a Royal Society of Chemistry journal dedicated to short communications in chemistry.

Activity 15.1 *Analyzing a Short Communication Paper*

Analyze the following short communication paper (from Kabatas Glowacki et al. 2022, *Chem. Commun.* 2022, **58**, 7558–7561) into the various sections discussed in Section 15.2, identifying the sentence structure and the verb tenses used in each section.

Activity 15.2 *Writing a Short Communication Paper*

Using the knowledge gained in this chapter, write a short communication paper for any completed simple experiment of your choice. This could be a routine or an occasional experiment in your laboratory.

REFERENCES

Kabatas Glowacki, S., P. Agüi-Gonzalez, S. Sograte-Idrissi, S. Jähne, F. Opazo, N.T.N. Phan, and S.O. Rizzoli. 2022. An iodine-containing probe as a tool for molecular detection in secondary ion mass spectrometry. *Chem. Commun.* 58:7558–7561.

Vezza, V.J., A. Butterworth, P. Lasserre, E.O. Blair, A. MacDonald, S. Hannah, C. Rinaldi, P.A. Hoskisson, A.C. Ward, A. Longmuir, S. Setford, E.C.W. Farmer, M.E. Murphy, and D.K. Corrigan. 2021. An electrochemical SARS-CoV-2 biosensor inspired by glucose test strip manufacturing processes. *Chem. Commun.* 57 (30):3704–3707.

16 Review Papers

16.1 FUNCTIONS OF A REVIEW PAPER

A review paper (or article) critically and constructively surveys published literature on a subject area through analysis, classification and comparison (Mayer 2009) with the ultimate aim of synthesizing and summarizing the information gained for a general conclusion or illustration of a point of view. A review paper relies solely on previously published literature but may also contain a small percentage of unpublished results. The audience of review papers include:

- Experts of the subject area.
- Students or newcomers to the subject area.
- Policy makers.

The functions of review papers are (Mayer 2009, Lau and Kuziemsky 2016, Palmatier et al. 2018) as follows:

- To give the historical background of a subject area.
- To give an overview of the current thinking of a subject area.
- To organize and evaluate published literature on a subject area.
- To serve as a lecture note or guide for students and newcomers to a subject area.
- To identify the most important literature on a subject area.
- To identify pattern and trends in published literature on a subject area.
- And/or to identify research gaps in a subject area and recommend potential research directions.

16.2 TYPES OF REVIEW PAPERS

It is difficult to classify review papers due to the overlapping characteristics of various review papers (Aveyard and Bradbury-Jones 2019). However, review papers can be classified loosely under three broad themes (Mayer 2009): (i) method of preparation, (ii) specific objective of the review and (iii) mandate of the review. In chemistry, the first class includes narrative, systematic and critical reviews while the second class includes historic, status quo, theory/model and issue reviews. The third class includes solicited and unsolicited reviews. A detailed classification of review papers can be found in Paré et al. (2015), Oosterwyk et al. (2019), Grant and Booth (2009) and Cooper (1988).

DOI: 10.1201/9781003186748-16

16.2.1 METHOD OF PREPARATION

a. *Narrative review*

- Gives a broad overview of a topic without a specific question. That is, a narrative review accumulates and summarizes or synthesizes published literature on a topic but does not generalize knowledge gained from the research. The author(s) simply accumulate(s) and synthesize(s) published literature to illustrate a given point of view (Callcut and Branson 2009).

- Literatures are selected, compared and summarized based on the experience of the author(s), available theories and models (Callcut and Branson 2009). Statistical analysis is not involved; as a result, the conclusions derived from the study are qualitative. The conclusions may also be intentionally or unintentionally biased by the expertise and the experience of the author(s) (Callcut and Branson 2009).

- *Functions* (Cronin et al. 2008, Green et al. 2006): (i) Gives the reader a comprehensive information and background of the topic to enable easy understanding of new knowledge. (ii) Brings practitioners of the field up to date with current trends. As such, narrative reviews are often used as lecture notes as they are more current than textbooks. (iii) Inspires new researches in the field by identifying areas that have received little attention. Narrative reviews are more common in chemistry compared with systematic reviews. It is common to have elements of *critical reviews* in *narrative reviews.*

- *Examples*: (i) "The role of particles in stabilizing emulsions and foams" (Hunter et al. 2008). (ii) "Photoelectrochemical devices for solar water splitting – materials and challenges" (Jiang et al. 2017). (iii) "Wearable and flexible electronics for continuous molecular monitoring" (Yang and Gao 2019).

b. *Critical review*

- Critically analyzes, evaluates and interprets published literature on a topic to expose strengths, weaknesses, inconsistencies, controversies and contradictions in relation to hypotheses, theories, models, methodologies or findings (Lau and Kuziemsky 2016, Toronto and Remington 2020). This does not imply the literature is criticized negatively. Criticizing is easy, but valueless. It is more useful to point out how further research can build on existing literature than pointing out flaws.

- *Functions*: (i) To assess the credibility of published literature on a topic as well as give a reflective account. (ii) To point out weaknesses and strengths, thereby giving direction for further research and improvements (Lau and Kuziemsky 2016). Critical reviews are also common in chemistry and may contain elements of narrative reviews.

- *Examples*: (i) "Contact angle hysteresis – advantages and disadvantages: A critical review" (Tyowua and Yiase 2020). (ii) "Wheat (*Triticum aestivum L.*) bran in bread making: A critical review" (Hemdane et al.

2016). (iii) "Nutritional chemistry of the peanut (*Arachis hypogaea*)" (Toomer 2018).

c. **Systematic review**

- Covers a narrow scope of a topic with a specific question to answer at the end of the review.
- Literature on a specific topic, within a given period of time, is selected from databases and analyzed statistically, provided the data is from the same source (homogeneous). The ultimate aim is to answer a specific question or reach a general conclusion (Callcut and Branson 2009).
- Systematic reviews may use meta-analysis which involves statistical combination of homogeneous data from two or more literatures or vote counting, to obtain an answer to a research question (Callcut and Branson 2009, Lau and Kuziemsky 2016). Systematic reviews are rather called "qualitative systematic reviews" when vote counting is used (Lau and Kuziemsky 2016). Unlike narrative reviews, the conclusions drawn from systematic reviews are more impactful and less prone to author bias (Callcut and Branson 2009).
- *Function*: A systematic review provides answer to a specific question on a topic just like a research paper addresses a specific research gap in a subject area. Systematic reviews are more common in medical and biomedical sciences (Ferrari 2015) and research in education and agriculture.
- *Examples*: (i) "Science education textbook research trends: A systematic literature review" (Vojíř and Rusek 2019). (ii) "Gender differences in mathematics and science competitions: A systematic review" (Steegh et al. 2019).

16.2.2 SPECIFIC OBJECTIVE OF REVIEW

a. **Historical review**

- Reviews the growth and development of a subject area within a given period of time (Noguchi 2006).
- *Function*: To give a historical account of the development of a subject area.
- *Examples*: (i) "The historical development of mechanochemistry" (Takacs 2013). (ii) "Early (pre-DLVO) studies of particle aggregation" (Vincent 2012). (iii) "History of lithium ion batteries" (Scrosati 2011).

b. **Status quo review**

- Reviews the current thinking and state-of-the-art research on a subject area (Noguchi 2006).
- *Function*: To describe the current thinking and state-of-the-art research on a subject area, pointing out areas that need further research attention.
- *Examples*: (i) "Current trends in the chemistry of permanent hair dyeing" (Morel and Christie 2011). (ii) "Basic research methods and current trends of dental implant surfaces" (Coelho et al. 2009).

c. *Theory/model review*
 • Reviews the development and introduction of a theory or model in a subject area (Noguchi 2006).
 • *Function*: To show the extent to which a theory or model fits experimental data on a subject area, pointing out successes and drawbacks as well as giving recommendations for future research.
 • *Examples*: (i) "Accounting for natural organic matter in aqueous chemical equilibrium models: A review of the theories and applications" (Dudal and Gérard 2004). (ii) "Application of chitosan for the removal of metals from wastewaters by adsorption – mechanisms and models review" (Gerente et al. 2007).

d. *Issue review*
 • Reviews a contentious issue in a subject area, giving a balanced view of both sides of the argument as well as a personal perspective (Noguchi 2006).
 • *Function*: To give an overview of a contentious issue, giving a fair attention to both sides of the argument as well as a reflective personal perspective.
 • *Examples*: (i) "Green chemistry and the health implications of nanoparticles" (Albrecht et al. 2006). (ii) "The chemistry of beer aging – a critical review" (Vanderhaegen et al. 2006).

16.2.3   MANDATE OF REVIEW

a. *Solicited review*
 • Also called (*featured*) *invited article* or commissioned review.
 • An experienced researcher on a subject area is invited by a journal editor to write a review on a specific topic (Mayer 2009).
 • *Function*: To increase the credibility of the journal and encourage other authors to submit related reviews and research papers on the topic.
 • *Examples*: (i) "Colloidal particles at a range of fluid-fluid interfaces" (Binks 2017). (ii) "Variations in meat color due to factors other than myoglobin chemistry; a synthesis of recent findings" (Purslow et al. 2020). (iii) "Where and how 3-D printing is used in teaching and education" (Ford and Minshall 2019).

b. *Unsolicited review*
 • A researcher develops a review paper without invitation (Mayer 2009).
 • *Functions*: (i) To critically analyze published literature and expose the current state of the subject area. (ii) To compare and contrast ideas on topical issues, stating any limitations. (ii) To identify contentious areas/ideas as well as research gaps. (iii) To recommend future research directions.
 • *Examples*: Almost all review articles published in chemistry journals dedicated to reviews like "Chemical Society Reviews", "Advances in Colloids and Interface Science" and "Chemical Reviews" are

unsolicited. These articles are written by recognized experts of various subject areas.

Activity 16.1 *Classifying Review Papers*

Read at least five review papers in your area of specialty and classify them based on the proceeding classes of review papers discussed.

16.3   WRITING A REVIEW PAPER

Writing a literature review involves two main steps (Pautasso 2013, Cronin et al. 2008). **Step 1** (*preparation*) involves: (a) defining a topic and audience; (b) obtaining relevant literatures; (c) reading, analyzing and critiquing or evaluating the literatures; and (d) synthesizing important information from the literatures. **Step 2** (*writing*) involves writing and referencing.

16.3.1   PREPARATION

a. ***Defining a topic and audience***
   Decide the type of review you want to go for and choose a suitable topic.
   - The topic should be interesting to your audience (*i.e.*, experts in the field and students or newcomers to the field).
   - To keep the review to a manageable limit, the topic should not be too wide.
   - The topic should have sufficient (> 50) literature to warrant a review.

b. ***Obtaining relevant literature***
   - Use multiple databases to search for literature on your topic so that you do not miss out any literature. Common databases include Google Scholar, Web of Science, Scopus, Toxline, ChemSpider, ChemRxiv and Mendeley. Boolean operators ("AND", "OR", "NOT") are quite useful for searching many databases (Ely and Scott 2007).
   - Keep track of your search terms to enable replication of your search.
   - To facilitate management of literature, obtain all literature on the topic, including previous review papers, into a paper management system like Mendeley, Qiqqa or Endnote. Also, decide a criterion for excluding irrelevant literatures from your collection and keep track of it to enable replication. For example, a conference paper which is subsequently published as a journal article can be excluded from the literature list.

c. ***Reading, analyzing and evaluating literature***
   - Read, analyze and evaluate your literature.
   - For each paper (Derish and Annesley 2011),
     – make a note of the
        (i)   aim or hypothesis and
        (ii) main finding(s)

      − examine (for *critical reviews*) the

          (iii) strengths and weakness or limitations of the findings (if any) and

          (iv) contribution of the work to the topic being reviewed,

      − evaluate (for *critical reviews*)

          (v) the suitability of the methods for the hypothesis tested and

          (vi) whether or not the interpretation of the results and conclusions drawn are sound.

d. *Information synthesis*

- Synthesize (*i.e.*, combine) related findings from various papers and make a summary. Use the summary as a framework of your review paper.

16.3.2 WRITING AND REFERENCING

A well-written review paper tells a story succinctly just like a well-written research paper. In both cases, the story consists of similar ingredients. For a research paper, the ingredients are a well-articulated (Derish and Annesley 2011) (i) knowledge gap, (ii) method for filling the gap, (iii) outcome(s) of experiment(s), (iv) meaning of the outcome(s), (v) conclusion(s) and recommendation(s) and (vi) importance of the work. Similarly, the ingredients for a review paper are a well-articulated (Derish and Annesley 2011) (i) reason for writing the paper, (ii) method through which literatures were obtained and reviewed, (iii) outcomes or findings of the literature as well as the meaning of the findings, (iv) conclusions and recommendations and (v) areas for further research.

- Write your review, using your summary notes as a framework and include all the relevant references.
- Ensure that all the ingredients listed here are clearly visible. To achieve this, think of your review as a press release. Say what is new, exciting and praise-worthy of your review compared with previous reviews. It could be that the topic has not been reviewed despite numerous published literatures or that much has changed – in terms of methodology, theories or models – since the last review of the topic. Perhaps, it could be that your review settles existing controversies (Derish and Annesley 2011). This will form the foundation of your review upon which the other ingredients can be added and built upon.

16.4 ELEMENTS OF A REVIEW PAPER

A review paper typically contains the following elements (*i.e.*, parts or sections): (a) a title, (b) list of authors and their affiliations, (c) a graphical abstract, (d) a textual abstract, (e) a video abstract, (f) keywords, (g) highlights, (h) an introduction, (i) a body and (j) a conclusion. The title, author list and address/affiliation, graphical abstract, textual abstract, video abstract, keywords and highlights of a review paper have the same function as those of a research paper and thus can be written using the guidelines given in Chapters 3–8. However, the introduction, body and conclusion sections of a review paper have slightly different functions compared with those of a research paper and should be prepared using the guidelines given in this chapter. The best section to start writing is the *body*, followed by the *conclusion*, the *introduction* and then the *abstracts*, respectively (Ferrari 2015). The rest of the sections can be written in any order. The general framework for writing a review paper is given in Table 16.1.

a. *Title*
   - The title should: (i) be short (maximum 12 words) and concise, (ii) contain keywords, (iii) indicate the type of review and (iv) be informative to enable readers to decide whether to read it or not.
   - The title can be written in either simple present or past tense or as a question. Use a simple present tense title to indicate that the information contained in the paper is a valid fact, use a simple past tense title to indicate that the information contained in the paper is not an established fact and use a question title to indicate that the question was open at the time of writing the review.

b. *Author(s) and author affiliation*
   - The author list should include everyone that has contributed in one way or the other to the paper.
   - Discuss and decide the order of names on the author list as soon as possible, perhaps at the planning stage, so as to avoid disagreements during the publication stage.
   - The authors' address is their organization at the time of the review. If an author has moved to another organization before publication of the paper, the name of their former and new organizations must be indicated as their address and current affiliation, respectively.

c. *Textual abstract, graphical abstract, video abstract and highlights*
   - The textual abstract (200–250 words) should be either *informative* or *descriptive*. Informative abstracts state the specific *objectives* and the main *outcomes* of the review while descriptive abstracts rather indicate the *contents* and *structure* of the review (Mayer 2009).
   - *Descriptive abstract*: A descriptive abstract is a table of contents written in a paragraph form. Descriptive abstracts are written in the simple present tense, and they are commonly used for narrative reviews.
   - *Informative abstract*: (i) Objectives – use one to two sentences (simple present tense) to state the context and intention of the review. (ii) Materials and methods (systematic reviews only) – use two to four sentences (simple past tense) to give the materials and the methodological approach used. (iii) Results – use two to four sentences (simple past tense) to describe the main outcomes of the review. (iv) Conclusions – use one to two sentences (simple present tense) to state the conclusions derived from the review (Mayer 2009).

TABLE 16.1

General Framework for Writing a Review Paper

Title	A precise and short title (max. 12 words)
Author(s)	One or more authors; author list should include those that have contributed significantly in one way or the other
Abstract	• Descriptive (narrative and critical reviews) • Informative (systematic reviews)
Keywords	Words or phrases that succinctly describe the paper and will enable its retrieval in online searches
Introduction	**Paragraph 1:** Background/context **Paragraph 2:** Problem **Paragraph 3:** Motivation or justification and structure of review **Paragraph 4** (narrative and critical reviews only): Literature search strategy (databases and keywords) Inclusion and exclusion criteria
Materials and methods (Systematic reviews only)	Data sources (*e.g.*, databases), search strategy (keywords and phrases), timeframe, inclusion and exclusion criteria and statistical methods (meta-analysis)
Body	**Section 1** (first key concept): Discuss and evaluate; summarize in relation to the aim of the review **Section 2** (second key concept): Discuss and evaluate; summarize in relation to the aim of the review **Section 3** (third key concept): Follow the same pattern
Conclusion(s)	• Summarize the main points per section and state the meaning in relation to the aim of the review • State the overall outcome and implication – *i.e.*, the lesson learned from the review • Direction for future research
Acknowledgments	Thank and appreciate all that have helped in one way or the other, including funding organizations
References	A comprehensive list of all the literatures used in the review

- The guidelines for preparing video abstracts, graphical abstracts and highlights of review papers are the same as those for research papers discussed in Chapters 5, 7 and 8, respectively.

d. **Keywords**
- Choose keywords or phrases that describe your paper exactly and that will enable its retrieval in online searches.

e. **Introduction**
- The introduction should contain at least three paragraphs that clearly state the following:
  - (i) Subject background/context – give the general topic, issue or area of concern.
  - (ii) Problem – state a single problem (an explicit question) or give trends, new perspective, gaps or conflicts. Be sure that you have a narrow focus.
  - (iii) Motivation or justification and structure – state the reason(s) for the review and give the structural organization of the paper.
- As much as possible, use the simple present tense to state facts and established knowledge and the simple past tense to describe unestablished knowledge.

f. **Body (narrative, critical and systematic reviews)**

Narrative reviews
- Narrative reviews tell a story about the present that shapes the future (Myers 1991). "The writer of the review shapes the literature of a field into a story in order to enlist the support of readers to continue that story" (Myers 1991).
- Divide the body of the review into suitable subsections which will serve as the *plot* of the story. Tell your story by (i) describing work and findings (major and minor) that have accrued over time, (ii) stating competing theories or models and (iii) stating further work that needs to be done.
- For each section, cover one idea per paragraph; however, the paragraph should not contain work from only a single literature. Instead, consider several literatures per paragraph.
- Ensure that there is a coherent relationship between the subsections. This can be achieved by stating (i) the link between each subsection and the specific objective(s) or review question and (ii) the relationship between the literatures.
- Use (Mayer 2009):
  - (i) The simple present tense to state general knowledge, established facts and what another author has reported or said, *e.g.*, "they show that …"
  - (ii) The present perfect tense to summarize what individual authors have reported on an issue in separate literatures, *e.g.*, "they have reported …"
  - (iii) The simple past tense to refer to a single study or to report what another researcher found, *e.g.*, "Taylor reported …"
- Use different referencing methods, *viz.* information prominent method, author prominent method or weak author prominent method (see Chapter 9, Section 9.3).

- Include literature sources (bibliographic databases), search terms and strategies, literature inclusion and exclusion criteria and the literature size to minimize bias and enable reproducibility as recommended by Ferrari (2015).

Critical reviews
- Critical reviews evaluate and challenge knowledge reported in literatures on a topic based on the authors' perception or point of view. This essentially involves questioning reported knowledge and presenting an evaluation or judgment of it.
- Divide the body of the review into suitable subsections that address only one specific idea. Each subsection can be structured as follows:
    - (i)   First idea: Explanation and evaluation.
    - (ii)   Second idea: Explanation and evaluation.
    - (iii) Third idea: Explanation and evaluation and so on.
- The evaluation should include both the strengths and the weaknesses or limitations of the literatures. However, negative criticisms should be avoided.
- Aim for a synthesized argument.
- A statement of how further research can build on the existing literatures can be added.

Systematic reviews (materials and methods)
- Systematic reviews have a materials and method section (written in the simple past tense) that clearly states:
    - (i)   The source of data (*e.g.*, bibliographic databases), search strategies and terms.
    - (ii)   Literature (or data) inclusion and exclusion criteria.
    - (iii) Size of data sample.
    - (iv)   Statistical methods of analysis.

g. ***Conclusion(s)***

Narrative reviews
- The concluding section should be written in simple present tense and should clearly articulate (Mayer 2009):
    - (i)   The main finding(s) derived from the literature review as well as the implications of the finding(s).
    - (ii)   The authors' interpretation of the findings.
    - (iii) List of unresolved issues.
    - (iv)   Directions for future research.

Critical reviews
- State your overall opinion as well as further explanation of your judgment, mentioning strengths and weaknesses.
- Make recommendations for future work.

Systematic reviews
- The concluding section should contain the answer to the research question posed in the introduction, the meaning and implications of the findings as well as the recommendations (Mayer 2009).

h. *Acknowledgments*
- Thank all the people that have helped in one way or the other, but are not on the author list. Write their full names and their specific contribution(s).
- Thank all the organizations that have funded the research. Write their full names, the program under which the work was funded as well as the grant numbers.

i. *References*
- Should contain a comprehensive list of all the literatures used in the review with an indication of the most important ones.

Activity 16.2 *Writing a Review Paper*

Write a 4000-word review paper on any aspect of your research area, indicating the type of review in the title.

FURTHER READING

Ridley, D. 2012. *The Literature Review: A Step-by-Step Guide for Students.* 2 ed. UK: Sage.
- Contains a comprehensive guide for writing a literature review. Detailed information on the planning stage, literature search, reading, note taking and writing stage are given.

Toronto, C.E., and R. Remington. 2020. *A Step-by-Step Guide to Conducting an Integrative Review*: Springer Nature, Switzerland.
- Contains a comprehensive guide for conducting and writing an integrative review.

REFERENCES

Albrecht, M.A., C.W. Evans, and C.L. Raston. 2006. Green chemistry and the health implications of nanoparticles. *Green Chem.* 8 (5):417–432.

Aveyard, H., and C. Bradbury-Jones. 2019. An analysis of current practices in undertaking literature reviews in nursing: Findings from a focused mapping review and synthesis. *BMC Med. Res. Methodol.* 19 (1):105.

Binks, B.P. 2017. Colloidal particles at a range of fluid–fluid interfaces. *Langmuir* 33 (28):6947–6963.

Callcut, R.A., and R.D. Branson. 2009. How to read a review paper. *Respir. Care* 54 (10):1379.

Coelho, P.G., J.M. Granjeiro, G.E. Romanos, M. Suzuki, N.R.F. Silva, G. Cardaropoli, V.P. Thompson, and J.E. Lemons. 2009. Basic research methods and current trends of dental implant surfaces. *J. Biomed. Mater. Res. B* 88B (2):579–596.

Cooper, H.M. 1988. Organizing knowledge syntheses: A taxonomy of literature reviews. *Knowledge in Society* 1 (1):104.

Cronin, P., F. Ryan, and M. Coughlan. 2008. Undertaking a literature review: A step-by-step approach. *Br. J. Nurs.* 17 (1):38–43.

Derish, P.A., and T.M. Annesley. 2011. How to write a rave review. *Clin. Chem.* 57 (3):388–391.

Dudal, Y., and F. Gérard. 2004. Accounting for natural organic matter in aqueous chemical equilibrium models: A review of the theories and applications. *Earth-Sci. Rev.* 66 (3):199–216.

Ely, C., and I. Scott. 2007. *Essential Study Skills for Nursing.* Edinburgh: Elsevier.

Ferrari, R. 2015. Writing narrative style literature reviews. *Medical Writing* 24 (4):230–235.

Ford, S., and T. Minshall. 2019. Invited review article: Where and how 3D printing is used in teaching and education. *Addit. Manuf.* 25:131–150.

Gerente, C., V.K.C. Lee, P.L. Cloirec, and G. McKay. 2007. Application of chitosan for the removal of metals from wastewaters by adsorption—mechanisms and models review. *Crit. Rev. Environ. Sci. Technol.* 37 (1):41–127.

Grant, M.J., and A. Booth. 2009. A typology of reviews: An analysis of 14 review types and associated methodologies. *Health Infor. Libr. J.* 26 (2):91–108.

Green, B.N., C.D. Johnson, and A. Adams. 2006. Writing narrative literature reviews for peer-reviewed journals: Secrets of the trade. *J. Chiropr. Med.* 5 (3):101–117.

Hemdane, S., P.J. Jacobs, E. Dornez, J. Verspreet, J.A. Delcour, and C.M. Courtin. 2016. Wheat (Triticum aestivum L.) bran in bread making: A critical review. *Compr. Rev. Food Sci. Food Saf.* 15 (1):28–42.

Hunter, T.N., R.J. Pugh, G.V. Franks, and G.J. Jameson. 2008. The role of particles in stabilising foams and emulsions. *Adv. Colloid Interface Sci.* 137 (2):57–81.

Jiang, C., S.J.A. Moniz, A. Wang, T. Zhang, and J. Tang. 2017. Photoelectrochemical devices for solar water splitting – materials and challenges. *Chem. Soc. Rev.* 46 (15):4645–4660.

Lau, F., and C. Kuziemsky. 2016. *Handbook of eHealth Evaluation: An Evidence-Based Approach.*

Mayer, P. 2009. Guidelines for writing a review article. *Zurich-Basel Plant Science Center: Zurich, Switzerland* 82:443–446.

Morel, O.J.X., and R.M. Christie. 2011. Current trends in the chemistry of permanent hair dyeing. *Chem. Rev.* 111 (4):2537–2561.

Myers, G. 1991. Stories and styles in two molecular biology review articles. *In Textural Dynamics of the Professions. Historical and Contemporary Studies of Writing in Professional Communities*, edited by C Bazerman and J Paradis, 45–75: University of Winconsin Press, Madison, WI.

Noguchi, J. 2006. *The Science Review Article: An Opportune Genre in the CONSTRUCTION OF SCIENCE. Linguistic Insights Volume 17.* Bern: Peter Lang.

Oosterwyk, G., I. Brown, and S. Geeling. 2019. A Synthesis of Literature Review Guidelines from Information Systems Journals. *Proceedings of 4th International Conference on the Internet, Cyber Security and Information Systems.*

Palmatier, R.W., M.B. Houston, and J. Hulland. 2018. Review articles: Purpose, process, and structure. *J. of the Acad. Mark. Sci.* 46 (1):1–5.

Paré, G., M.-C. Trudel, M. Jaana, and S. Kitsiou. 2015. Synthesizing information systems knowledge: A typology of literature reviews. *Infor. Manag.* 52 (2):183–199.

Pautasso, M. 2013. Ten simple rules for writing a literature review. *PLoS Comput. Biol.* 9 (7):e1003149.

Purslow, P.P., R.D. Warner, F.M. Clarke, and J.M. Hughes. 2020. Variations in meat colour due to factors other than myoglobin chemistry: A synthesis of recent findings (invited review). *Meat Sci.* 159: 107941.

Scrosati, B. 2011. History of lithium batteries. *J. Solid State Electrochem.* 15 (7):1623–1630.

Steegh, A.M., T.N. Höffler, M.M. Keller, and I. Parchmann. 2019. Gender differences in mathematics and science competitions: A systematic review. *J. Res. Sci. Teach.* 56 (10):1431–1460.

Takacs, L. 2013. The historical development of mechanochemistry. *Chem. Soc. Rev.* 42 (18):7649–7659.

Toomer, O.T. 2018. Nutritional chemistry of the peanut (Arachis hypogaea). *Crit. Rev. Food. Sci. Nutr.* 58 (17):3042–3053.

Toronto, C.E., and R. Remington. 2020. *A Step-by-Step Guide to Conducting an Integrative Review*: Springer.

Tyowua, A.T., and S.G. Yiase. 2020. Contact angle hysteresis - advantages and disadvantages: A critical review. *Rev. Adhesion Adhesives* 8: 47–67.

Vanderhaegen, B., H. Neven, H. Verachtert, and G. Derdelinckx. 2006. The chemistry of beer aging – a critical review. *Food Chem.* 95 (3):357–381.

Vincent, B. 2012. Early (pre-DLVO) studies of particle aggregation. *Adv. Colloid Interface Sci.* 170 (1):56–67.

Vojíř, K., and M. Rusek. 2019. Science education textbook research trends: A systematic literature review. *Int. J. Sci. Educ.* 41 (11):1496–1516.

Yang, Y., and W. Gao. 2019. Wearable and flexible electronics for continuous molecular monitoring. *Chem. Soc. Rev.* 48 (6):1465–1491.

17 Perspective Papers

17.1 FUNCTIONS OF A PERSPECTIVE PAPER

As the name implies, perspective papers are written by experts of a subject area based on their personal perspectives. They are very similar to review papers in terms of structure and length (≤ 7000 words, excluding figures, tables and references), but they are very opinionated because they are based on the personal opinion(s) of the author(s). Authors of perspective papers are expected to reflect deeply on a subject area and then write a stimulating essay (*i.e.*, the perspective paper) on the area based on their reflection.

A perspective paper is ultimately meant to (i) address a contentious issue, (ii) dismantle old ideas and put forward new ones, (iii) identify research gaps and (iii) give a visionary future research direction. However, all of these are based on the reflective personal opinion(s) of the author(s). Referees of perspective papers are asked to review them on the merits of quality and relevance of arguments irrespective of whether they agree with the opinions of the author(s) or not. Like review papers, perspective papers contain a title, author names and their affiliations, abstracts (textual, graphical, video), a list of keywords, an introduction, a body and an opinion section (an important and unique feature), conclusion(s) and references.

17.2 WRITING A PERSPECTIVE PAPER

Editors of journals often invite expert authors and provide them with guidelines to write perspective papers for an upcoming volume. Sometimes, expert authors also request permission from editors to write perspective papers on topics that are pertinent to the authors. Either way, an outline of the paper is agreed by both the editor and the authors before writing commences. All the guidelines given in Chapter 16 for writing the various sections of a review paper are applicable to the corresponding sections of a perspective paper, thus will not be duplicated here. Be sure to follow them when writing a perspective paper. However, I would like to draw your attention to the "opinion section" which is a unique feature of perspective papers. This is the section that makes perspective papers what they are and must be given great attention. This section should contain the following points:

- Your personal interpretation of the literature surveyed.
- Problems and unanswered questions.
- Future developments that are likely to be important to the area.
- Possible research gaps and directions.
- New ideas that will advance the area.
- A visionary, speculative and an evolutionary view of the area with time (say in the next 5–10 years).

DOI: 10.1201/9781003186748-17

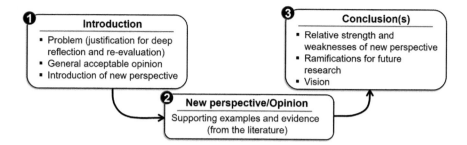

FIGURE 17.1   Illustration of the U-shaped structure of a perspective paper where the introduction section occupies the left arm, the perspective section occupies the base, while the conclusion section occupies the right arm.

However, in the absence of an opinion section, these points are scattered in the paper as can be seen in Whittingham (2020). With an opinion section, a perspective paper has a U-shaped structure (Figure 17.1) when the section is coupled with the introduction and the conclusion sections. This structure begins with a clear justification of why the topic needs a deep reflection and a re-evaluation and ends with a new personal perspective of the topic, potential implications of the new perspective and a vision for the topic.

Examples of perspective papers can be found in the following journals: "Nature Review Chemistry", "Organic and Biomolecular Chemistry" and "Frontiers in Chemistry" all of which publish high-quality and excellent perspective papers.

REFERENCE

Whittingham, M.S. 2020. Special editorial perspective: Beyond Li-Ion battery chemistry. *Chem. Rev.* 120 (14):6328–6330.

18 Cover Letters

18.1 FUNCTIONS OF A COVER LETTER

Many chemistry journals demand that a submitted manuscript should be accompanied by a cover letter. However, only few of them are explicit on the contents of the said cover letter. The cover letter is written, by the corresponding author, specifically for the editor-in-chief or editor, and its functions are to emphasize the importance of the findings reported and to also state that the manuscript (Moustafa 2015):

- Is original and it is not being considered in another journal for publication.
- Has been prepared in line with the journal's specifications.
- Is within the scope of the journal and that the journal is the best place to publish it.

The cover letter is also an opportunity to show the editor-in-chief or editor that you value their role in improving the manuscript and getting it published.

Despite these functions, the role of cover letters in scientific publishing has been questionable as their key contents (key findings, novelty of work and implications) are mandatorily contained in the abstract and the conclusion sections of the submitted paper. In other words, cover letters are criticized for having a redundant role in scientific publishing (Moustafa 2015). In fact, some editors-in-chief and editors do not take cover letters seriously (John 2011). However, those that take the cover letter seriously use it to a great effect, *e.g.*, for deciding whether or not a manuscript will progress to the peer-review stage (Holmes et al. 2009 and Hafner 2010). Therefore, a carefully planned, written and persuasive cover letter will improve a manuscript's chances of being selected for peer-review (Volmer and Stokes 2016, Hafner 2010) and vice versa for a poorly written and an unpersuasive cover letter (Hafner 2010).

18.2 WRITING A COVER LETTER FOR RESEARCH AND COMMUNICATION PAPERS

Because research and short communication papers have the same structure, their cover letters contain the same information. The cover letter is short (≤ 200 words) (John 2011), concise, written on a letter-headed paper and contains the following information (Cargill and O'Connor 2013, Holmes et al. 2009, Volmer and Stokes 2016 and John 2011):

DOI: 10.1201/9781003186748-18

- The corresponding author's name, address as well as contact information (*e.g.*, email and phone number). The letter is usually addressed to the editor-in-chief or editor (to whom the letter is written); either the institutional address of the editor or journal address is acceptable.
- The title of the manuscript, the manuscript type (research paper or communication) and the names of all the authors as well as the date of submission.
- A statement that the manuscript has been prepared according to the journal's specifications.
- A brief background of the research question.
- The research question, highlighting how the work fits into the scope of the journal as well as how it will appeal to the journal's readership.
- Highlights of key findings.
- A statement that the research is original and it is not simultaneously considered elsewhere for publication.
- Reasons for certain features of the manuscript that might be questioned by the editor-in-chief or editor, *e.g.*, a justified reason for violating manuscript length.
- An ethical statement for research involving biological substances.
- Reason(s) for submitting the manuscript to the journal.
- (optional) Declaration of conflict of interest, names, addresses and emails of potential reviewers and an agreement statement saying "all the authors have read and agreed to submitting the manuscript to the journal".

After incorporating these pieces of information in your cover letter, end the cover letter saying "you look forward to the editor-in-chief's or editor's comments as well as those of the reviewers". This makes them feel that you value their role in improving the manuscript and getting it published. You can now sign the letter and submit it with the manuscript. Authors commonly make the mistake of copying the abstract into the cover letter – please avoid this! A typical structure of a research or short communication paper cover letter is given in Table 18.1 and illustrated in Example 18.1. Some journals provide a cover letter template which authors must follow (for instance, Example 18.2 from Taylor and Francis), but such templates generally contain the information given in Table 18.1. However, you should ensure to download the template and follow it.

TABLE 18.1

A Typical Structure of a Research and a Short Communication Paper Cover Letter

Component	Content
Header	Editor-in-chief's or editor's name and address
	Date of manuscript submission
Opening salutation	Dear Editor-in-chief or Editor/Specific name, *e.g.*, James
Body	
Paragraph I	Manuscript title and type
	List of authors
(manuscript's details)	Statement of whether or not the manuscript is invited or belongs to a special issue
	Statement of the manuscript's length: number of words, tables and figures (optional)
Paragraph II	Background of research reported
	Originality of research reported
	Key findings and their implications
	Suitability of journal for publication
	Statement about broader appeal to readers' interest
Paragraph III	Conflict of interest statement
(optional)	List of funding sources
	Ethical standard statement
	Authors submission approval
	List of potential reviewers
Footer	
Closing salutation	Sincerely; Faithfully; Yours sincerely or faithfully
	Signature of corresponding author
	Contact details of corresponding author

Example 18.1 *A Sample Cover Letter of a Research Paper*

Cover letter (p. 142) accompanying the submission of "Foaming Honey – Particle or Molecular Foaming Agent?" by Tyowua et al. (2020) to the *Journal of Dispersion Science and Technology.*

BENUE STATE UNIVERSITY, MAKURDI, NIGERIA
The Department of Chemistry

19th May 2020

Professor Orlando Rojas
Canada Excellence Research Chair
Chemical & Biological Engineering,
The University of British Columbia

Dear Rojas

This letter accompanies the submission of our manuscript:

Title Foaming Honey – Particle or Molecular Foaming Agent?

Authors Andrew T. Tyowua, Adebukola M. Echendu, Stephen G. Yiase,
Sylvester O. Adejo, Luter Leke, Emmanuel M. Mbaawuaga and Bernard P. Binks

Liquid foams find numerous applications in the cosmetic, the pharmaceutical and the food industries. These applications are impeded by the separation of the gaseous and the liquid phases. However, while the conditions for obtaining relatively stable aqueous and non-aqueous foams are well-known including the choice of foaming agents, those for obtaining stable honey foam are unknown because honey foam is yet to be studied. Honey foam will find various applications in the food, the cosmetic and the pharmaceutical industries. Thus, we have prepared honey foam using particle (calcium carbonate) or molecular (sodium lauryl sulfate) foaming agent, commonly used for foaming aqueous and non-aqueous systems for the first time. The ultimate aim is to determine which foaming agent is better in terms of foam volume and stability. Sodium lauryl sulfate produced more foam than calcium carbonate particles and the foam was stable for over four months compared with the foam from calcium carbonate particles which collapsed completely within four weeks. This is contrarily to what is known in aqueous and non-aqueous systems where particles produce quite stable foams compared with surfactant molecules.

This finding will be important to the food, the cosmetic and the pharmaceutical industries where foams are used for various applications. It will also be important to colloid scientists for the creation of novel honey foam-based products. On this basis, we believe that the work is suitable for the wide readership of the Journal of Dispersion Science and Technology and therefore suitable for publication therein as a full-length paper. This work is not currently considered for publication in another journal and all authors have approved the final version and its submission to your journal.

We look forward to your comments and those of the reviewers.

Yours sincerely

Andrew Tyowua

Andrew T Tyowua
BSc (BSU) PhD (Hull) MCSN MACS MRSC MICCON
Applied Colloid Science & Cosmeceutical Group

Tel +23480918XXXXX
Email atyowua@bsum.edu.ng

Activity 18.1 *Analyzing a Research Paper Cover Letter*

Analyze the cover letter in Example 18.1 into the various pieces of information described in Section 18.2. You will notice that the cover letter does not follow the exact order given in Table 18.1, but it contains the most essential information the editor-in-chief or editor will like to have about the manuscript.

Example 18.2 *A Research Paper Cover Letter Template from Taylor and Francis*

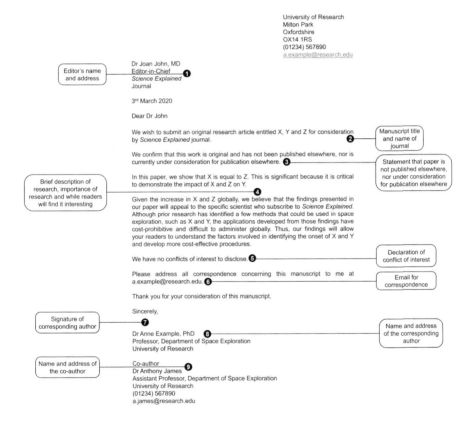

18.3   WRITING A COVER LETTER FOR REVIEW AND PERSPECTIVE PAPERS

The cover letters of review and perspective papers contain the same information as those of research and short communication papers. However, the background of the research and the research question are replaced by the background of the review and the aim of the review, respectively. Also, the research findings are replaced by the conclusions drawn from the review or personal perspective of the authors. Another difference between the cover letters of research papers and those of review and perspective papers is that the former might contain an ethical statement while the latter does not contain an ethical statement.

REFERENCES

Cargill, M., and P. O'Connor. 2013. *Writing Scientific Research Articles: Strategy and Steps*: John Wiley & Sons, New York.

Hafner, J.H. 2010. The art of the cover letter. *ACS Nano* 4 (5):2487–2487.

Holmes, D.R., P.K. Hodgson, R.A. Nishimura, and R.D. Simari. 2009. Manuscript preparation and publication. *Circulation* 120 (10):906–913.

John, M. 2011. I really think you should publish this paper: the cover letter to the editor. *HSR Proc. Intensive Care Cardiovasc. Anesth.* 3 (2):137.

Moustafa, K. 2015. Does the Cover Letter Really Matter? *Sci. Eng. Ethics* 21 (4):839–841.

Tyowua, A.T., A.M. Echendu, S.G. Yiase, S.O. Adejo, L. Leke, E.M. Mbawuaga, and B.P. Binks. 2020. Foaming honey: Particle or molecular foaming agent? *J. Dispersion Sci. Technol.* 43 (6):848–858.

Volmer, D.A., and C.S. Stokes. 2016. How to prepare a manuscript fit-for-purpose for submission and avoid getting a 'desk-reject'. *Rapid Commun. Mass Spectrom.* 30 (24):2573–2576.

19 The Publishing Process

19.1 PROCESSES PUBLISHED PAPERS UNDERGO

All published papers undergo three essential processes (Mizzaro 2003):

i. Submission – the manuscript is submitted to a target journal (*i.e.*, the journal it was planned and written for).
ii. Peer-review – the manuscript is appraised by experts for worthiness.
iii. Publication – the manuscript is published (if found worthy).

The breakdown of these processes and their relationship are summarized in Figure 19.1.

19.2 THE SUBMISSION PROCESS

The submission of manuscripts to many journals is now electronic *via* an online "submission management system" (*e.g.*, editorial manager). Ideally, the corresponding author is responsible for submitting the manuscript; however, the manuscript can also be submitted by a co-author on behalf of the corresponding author. Instructions and help for successfully submitting a manuscript are normally provided online at each stage of the submission process. Before submitting the manuscript, double-check:

• The *Guide for Authors* to ensure that all the instructions have been followed.
• That all the files required to complete the submission process are ready.

With all the files being ready and well-prepared in line with the *Guide for Authors*, submit the manuscript using the step-by-step submission instructions. The journal will send you an email, confirming your submission.

19.3 THE PEER-REVIEW PROCESS

The structure of the editorial team and the responsibilities of members of the team differ from journal to journal (Hill 2006). Four editorial team structures are common with many scientific journals; their composition, listed in the order of decreasing responsibilities and involvement, include (Hill 2006):

• **Structure 1** – *editor-in-chief, subject* (*associate*) *editors* or *editors, editorial board* and *international advisory board.*
• **Structure 2** – *editors* and *editorial board.*
• **Structure 3** – *editors, editorial board* and *international advisory board.*
• **Structure 4** – *editors, advisory board* and *editorial panel.*

DOI: 10.1201/9781003186748-19

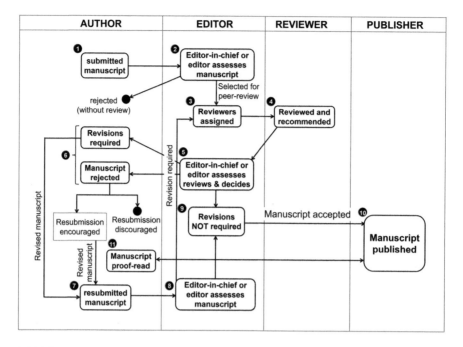

FIGURE 19.1 A flowchart showing details of the three main processes (submission, peer-review and publication) published manuscripts undergo before publication in a peer-reviewed journal. These processes are indicated by the key players, namely the author, the editor and the reviewer.

Irrespective of the structure, the ultimate goal is to have a journal that meets its set goals and also satisfies the expectations of its authors and readers. Authors expect a transparent and fair editorial process while readers expect quality papers that advance knowledge in the subject areas of the journal. Both of these are driven by an efficient editorial process. Regardless of the editorial structure, the editorial process is similar, with the key players being the editor-in-chief, editors (subject or associate) and editorial board (panel), all of whom are supposedly leading experts in the areas covered by the journal (Bedeian et al. 2008). The role of these players in the editorial process vary from journal to journal; but generally (Derntl 2014):

i. The *editor-in-chief* makes the final decision of accepting or rejecting a submitted manuscript in line with an editor's recommendation which is based on peer-review comments.

ii. The *editors* assign manuscripts to reviewers for peer-review, and they also serve as mediators between the corresponding author, reviewers and editor-in-chief.

iii. The *editorial board* reviews submitted manuscripts and helps editors in decisions making, *e.g.*, on plagiarism reports and manuscript selection when reviewers do not agree.

Depending on the journal, the *international advisory* or *advisory board* does not participate in manuscript appraisal; rather, it helps the editorial team in achieving the set goals of the journal. In order to satisfy the expectations of readers, journals select papers for publication using (i) editorial assessment and (ii) peer-review (Derntl 2014).

19.3.1 EDITORIAL ASSESSMENT

The editor-in-chief or editor checks the submitted manuscript for (Derntl 2014):

   i. Adherence to the instructions given in the *Guide for Authors*.
  ii. Agreement of the scope of the manuscript with the scope of the journal.
 iii. Originality of work – is the work or part of it published elsewhere?
  iv. Quality of writing and English language quality of the paper.
   v. Significance and contribution to the knowledge base of the subject area.

If the manuscript passes all these checks; it is selected and sent out for peer-review, but it is rejected without peer-review (Figure 19.1) if it fails any of these checks. This is known as *desk rejection*.

19.3.2 PEER-REVIEW

Peer-review is the quality-control process of scholarly publishing (Grainger 2007), and it involves the appraisal of a submitted manuscript by recognized experts (also known as "peers" or "reviewers") in the field of the manuscript. The time taken for peer-review varies from two weeks to several months. There are several types of peer-review systems, namely *open, single-blind, double-blind, triple-blind* and *quadruple-blind* systems (Haffar et al. 2019), with various merits and demerits (Haffar et al. 2019, Kelly et al. 2014).

   i. *Open peer-review system*: The identities of the authors, reviewers and editors are disclosed to one another.
  ii. *Single-blind peer-review system*: The identity of the authors is disclosed to the reviewers while that of the reviewers is undisclosed to the authors, but the editor knows the identities of both of them. The authors only know the identity of the editor.
 iii. *Double-blind peer-review system*: The identities of the authors and the reviewers are undisclosed to each but are both known to the editor. However, the identities of the editors and the authors are known to each other.
  iv. *Triple-blind peer-review system*: The identity of the authors is undisclosed to both the editor and the reviewers, but the identities of the reviewers and the editor are known to each other.
   v. *Quadruple-blind peer-review system*: The identities of the authors, editors and reviewers are undisclosed to one another throughout the peer-review and publishing processes.

Other systems of peer-review are also available (Ali and Watson 2016). Of these systems, single-blind peer-review is predominantly used by scientific journals (Ware 2008). However; whichever peer-review system is adopted by a journal, manuscript appraisal is based on the following criteria (Derntl 2014):

i. Originality – whether the research is original or it has been published (wholly or in parts) elsewhere.
ii. Clarity of writing – whether the paper is written in clear technical terms, well-organized, concise and readily understandable or not.
iii. Appropriateness of methods in relation to the research question and reproducibility of results – whether the methods are suitable for investigating the research question or not and whether the results obtained are reliable and can be reproduced or not.
iv. Appropriateness of title and abstract – whether the title and the abstract clearly describe the content of the paper or not.
v. Appropriateness of the discussion of results *vis-á-vis* the research question – whether the results obtained from the study are well-discussed in line with the research question or not.
vi. Appropriateness of the conclusions – whether the conclusions drawn from the study are in line with the research question and are well-supported by the results obtained.
v. The significance and contribution of the work to the body of knowledge of the field – whether the significance of the work is stated or not and whether the work advances knowledge of the field or not.

These criteria vary in importance depending on the journal, but the process leads to either of the three recommendations:

• Accept as it is (very rare) – the manuscript is published without any revisions.
• Revisions required (common) – the manuscript requires revisions (minor or major) before a publication is warranted.
• Or reject (very common) – the manuscript does not warrant publication in the journal at all.

The reviewers (2–5) send their comments as well as their recommendations to the editor-in-chief or editor who assesses them, decides and informs the corresponding author whether the manuscript (i) is to be published without revision, (ii) requires a revision before publication or (iii) does not merit publication (Figure 19.1). An example of typical peer-review comments can be found in Anonymous (2016a, b) for Marshall et al. (2016). Tips for responding to peer-review comments are given in Section 19.3.3. Sometimes, editors find it difficult to make a decision using reviewers' comments, especially when the comments do not agree. So, they use the editorial board or additional comments from independent reviewers to make the final decision. Although peer-review has been criticized for biasness, transparency and other issues (Ahmed and Gasparyan 2013, Wicherts 2016, Goldbeck-Wood 1999,

Lee et al. 2013), with proposed alternatives (Mizzaro 2003, Nentwich 2005, Correia and Teixeira 2005), it is still the quality-control process of scholarly publishing.

19.3.3 Responding to Peer-Review Comments

Responding to peer-review comments can be as difficult as writing the original paper. Nonetheless, how you respond to these comments often influences the final decision of whether or not to publish the paper. Authors often find peer-review comments unpleasant or irrelevant, but this is no excuse for arrogance and impolite responses. View each comment as a call to improve your manuscript. In light of this, Annesley (2011) has given ten tips for responding to peer-review comments, some of which are summarized here.

a. If the manuscript is rejected without the opportunity to resubmit, accept the decision of the editor, improve the manuscript using the comments and resubmit to another journal.
b. If you were given the opportunity to revise and resubmit the manuscript:
- Do not give argumentative and confrontational responses.
- Do not give arrogant and impolite responses.
- Peer-review comments may be asking for (i) clarification of text, (ii) further analysis, (iii) additional proof or references or (iv) additional experiments; be sure to provide responses to all the comments and give reasons for any experiments or analysis that you may have been unable to do.
- Even if a comment is not pertinent, do not respond by stating how the reviewer is wrong; rather, think of how you can improve the manuscript satisfactorily based on the comment.
- For opposing comments, decide what will improve the manuscript and explain your decision to the editor.
- Thank reviewers for any complimentary comments about the manuscript.
- Respond to the comments point by point (Appendix II), *i.e.*, copy the comments into your responses and respond immediately below.

19.4 PUBLICATION OF MANUSCRIPT

Following resubmission of a revised manuscript, reviewers might request for further revisions before accepting the manuscript for publication. Whatever is the case, once accepted, the manuscript is sent to the publishing manager or editor for publishing which often involves:

- *Copy-editing* where the manuscript's text, figures and tables are copied into the journal's publishing template and edited.
- *Proof-reading* where the authors receive the so-called "proofs" and correct any mistakes that might have occurred during copy-editing.

- *Signing of paper work* where the corresponding author signs the publishing agreement (on behalf of the co-authors), authorizing the publisher to publish the manuscript.
- *Deciding between the publishing options* (if the journal is not open access) – (i) "open access" where the paper is published at a fee but is freely and immediately available for all readers to download; (ii) "journal user license" where the paper is published for free, but users pay a fee before accessing it; or (iii) "non-commercial user license", *e.g.*, creative commons, where the paper is freely available for users to read, download, translate and reuse. Authors are normally encouraged to publish their papers under the open access option for the benefits it offers, but predatory open access journals and publishers are corrupting open access publishing (Beall 2012).

Congratulations! At this point your manuscript may be published online, ready to go into the next volume or issue of the journal.

FURTHER READING

Hames, I. 2008. Peer review and manuscript management in scientific journals: Guidelines for good practice: Blackwell Publishing, Oxford.
- Contains a comprehensive description of the peer-reviewing process as well as useful suggestions for reviewing scientific manuscripts.

Spyns, P., and M.-E. Vidal. 2015. *Scientific Peer Reviewing: Practical Hints and Best Practices*: Springer, Switzerland.
- Explains the peer-reviewing process and also contains useful suggestions for reviewing scientific manuscripts.

Responding to peer reviewer comments: A free example letter from Proof-reading-service.com
- Contains tips for responding to peer-review comments and an exemplary response to peer-review comments.

REFERENCES

Ahmed, H.S., and A.Y. Gasparyan. 2013. Criticism of peer review and ways to improve it. *Eur. Sci. Ed.* 39 (1):8–10.

Ali, P.A., and R. Watson. 2016. Peer review and the publication process. *Nurs. Open* 3 (4):193–202.

Annesley, T.M. 2011. Top 10 tips for responding to reviewer and editor comments. *Clin. Chem.* 57 (4):551–554.

Anonymous. 2016a. Peer review report 1 On "Hyperspectral narrowband and multispectral broadband indices for remote sensing of crop evapotranspiration and its components (transpiration and soil evaporation)". *Agric. For. Meteorol.* 217 (1):98–99.

Anonymous. 2016b. Peer review report 2 On "Hyperspectral narrowband and multispectral broadband indices for remote sensing of crop evapotranspiration and its components (transpiration and soil evaporation)". *Agric. For. Meteorol.* 217 (1):146.

Beall, J. 2012. Predatory publishers are corrupting open access. *Nature* 489 (7415):179–179.

Bedeian, A.G., D.D. Van Fleet, and H.H. Hyman. 2008. Scientific achievement and editorial board membership. *Organ. Res. Methods* 12 (2):211–238.

Correia, A.M.R., and J.C. Teixeira. 2005. Reforming scholarly publishing and knowledge communication: From the advent of the scholarly journal to the challenges of Open Access. *Inf. Serv. Use* 25 (1):13–21.

Derntl, M. 2014. Basics of research paper writing and publishing. *Int. J. Technol. Enhanced Learning* 6 (2):105–123.

Goldbeck-Wood, S. 1999. Evidence on peer review—scientific quality control or smoke-screen? *BMJ* 318 (7175):44.

Grainger, D.W. 2007. Peer review as professional responsibility: A quality control system only as good as the participants. *Biomaterials* 28 (34):5199–5203.

Haffar, S., F. Bazerbachi, and M.H. Murad. 2019. Peer review bias: A critical review. *Mayo Clin. Proc.* 94 (4):670–676.

Hill, M. 2006. The editorial board. In The E-Resources Management Handbook., edited by G Stone, R Anderson and J Feinstein: UK Serials Group, Newbury.

Kelly, J., T. Sadeghieh, and K. Adeli. 2014. Peer review in scientific publications: Benefits, critiques, & a survival guide. *EJIFCC* 25 (3):227–243.

Lee, C.J., C.R. Sugimoto, G. Zhang, and B. Cronin. 2013. Bias in peer review. *J. Am. Soc. Inf. Sci. Tec.* 64 (1):2–17.

Marshall, M., P. Thenkabail, T. Biggs, and K. Post. 2016. Hyperspectral narrowband and multispectral broadband indices for remote sensing of crop evapotranspiration and its components (transpiration and soil evaporation). *Agric. For. Meteorol.* 218–219:122–134.

Mizzaro, S. 2003. Quality control in scholarly publishing: A new proposal. *J. Am. Soc. Inf. Sci.* 54 (11):989–1005.

Nentwich, M. 2005. Quality control in academic publishing: Challenges in the age of cyber-science. *Poiesis Prax.* 3 (3):181–198.

Ware, M. 2008. *Peer Review: Benefits, Perceptions and Alternatives*. London: Publishing Research Consortium.

Wicherts, J.M. 2016. Peer review quality and transparency of the peer-review process in open access and subscription journals. *PLoS One* 11 (1):e0147913.

20 Poster Presentation

20.1 SCIENTIFIC POSTERS

A scientific poster is a one-page visual description of a research finding using a combination of text, sketches, illustrations and graphs. Akin to a graphical abstract, a scientific poster can also be thought of as a one-page illustrative abstract of a research finding. Posters are printed on either satin finish photo paper (photographic purpose), canvas (traveling purpose) or exhibition vinyl (exhibition purpose). Posters are supposed to be clearly understood by readers in 35 min (Block 1996). The standard print size of a scientific poster is A0 (841 × 1189 mm). However, depending on the purpose, posters can also be printed on A1 (594 × 841 mm) or A2 (420 × 594 mm) as well as any size (up to 15 m width and 1250 mm length). Posters were introduced in scientific meetings (Biophysical Chemists and the Biophysical Society) in the United States of America in 1974 (Hess et al. 2009) because of the limited time available for oral presentations. Now as an integral part of scientific conferences and meetings, posters can also be used for teaching (Hess and Brooks 1998) and assessment (Moneyham et al. 1996, Bracher 1998, Costa 2001). Furthermore, teachers can use posters to teach presentation and communication skills (Pelletier 1993) and to promote team work as well as instill critical thinking and analysis in students (Moule et al. 1998). Posters can also be used to assess students in place of the conventional written essays and exams (*i.e.*, both written and oral) (Akister et al. 2000).

There are thousands of online and offline materials with excellent suggestions on how to prepare a scientific poster. However, I have observed that many of the exemplary posters used in these materials are misleading and many of their suggestions are outdated and inconsistent. Thus, these suggestions cannot be used to create a poster that performs its sole function of disseminating scientific findings to audience in 3–5 min. In fact, only ~0.5% of posters out there achieve this important function. Therefore, this chapter gives useful and practical suggestions for preparing an effective and an appealing scientific poster for conferences and meetings. These suggestions can also be used to prepare posters for other academic purposes, like student assessment and exam, keeping in mind the reader/audience (Sousa and Clark 2019).

20.2 FEATURES OF AN EFFECTIVE AND AN APPEALING SCIENTIFIC POSTER

An effective scientific poster (Sousa and Clark 2019):

- Contains a single (relevant) message.
- Uses infographic techniques, *i.e.*, more illustrations (graphs, sketches and photos) with little text, to convey its message.
- Contains a well-ordered and obvious sequence of information.

DOI: 10.1201/9781003186748-20

- Speaks for itself, *i.e.*, it is well-understood without the presenter explaining.
- Is legible at a distance of 1–3 m.
- Is straight to the point.
- Passes the message across readily (say 3–5 min).

An appealing scientific poster (Sousa and Clark 2019):

- Is attractive (in terms of layout design).
- Contains the right font type and text size.
- Has a well-harmonized color mix.
- Contains carefully crafted illustrations (graphs and sketches).

These features are akin to those of an advertisement billboard or poster (Figure 20.1), and they indicate that poster making requires a combination of scientific and art creativity (Mitrany 2005, Chopra and Kakar 2014, Nundy et al. 2022).

The McDonald's soft drink advertisement billboard (Figure 20.1) simply and boldly says "$1 any size soft drink" (*i.e.*, $1 size fits all), with the key parts ("$1", "any size" and "all") boldfaced and photos of the soft drinks boldly displayed. With little text, appropriate photo, right text size and color combination, the billboard message is absolutely unequivocal. Like the advertisement billboard, authors of posters are in effect advertising or "selling" their work (*i.e.*, findings, argument, *etc.*) in the most attractive manner (Rowe 2017). Therefore, recognizing the advertising role of a scientific poster is the first step toward creating an effective and appealing poster. At conferences and meetings, these billboard-like features of an effective and appealing poster, outlined here, form the basis for judging posters for prize award. Coupled with these features, the final award decisions are made by taking cognizance of the author's interaction with the audience during presentation (Rowe 2017). Therefore, authors are encouraged to use these features to self-assess their posters before presenting them at scientific conferences and meetings.

FIGURE 20.1   Photograph of a McDonald advertisement billboard for the following soft drinks: Dr Pepper, Coca-Cola, Diet Coke and Sprite, saying with $1 you can get any size of the soft drinks.

20.3   SOFTWARE PACKAGES FOR PREPARING SCIENTIFIC POSTERS

There are plenty of software packages that can be used to prepare scientific posters, some of which include:

- Microsoft PowerPoint (commonest, easy to use, useful for typography, photo editing; creates relatively light files).
- Microsoft Publisher (professional, suitable for typesetting, page layout and design).
- CorelDraw (professional, useful for typography, photo editing and vector illustration, but creates relatively heavy files).
- Adobe Freehand (professional, useful for creating vector graphics).
- OmniGraffle (professional, useful for drawing, vector graphics and digital illustration).
- Inkscape (professional and free, useful for creating vector imagery, typography, compatible with a broad range of file format).
- QuarkXPress (professional, useful for graphic designs and photo editing, but it is relatively expensive).
- PosterGenius (mainly for designing posters but has a steep learning curve).
- Scribus (professional and free, suitable for layout, typography, image editing, publishing as well as animation).
- Affinity Designer (professional, suitable for illustration, vector graphics and digital art design).
- Google slides (web-based, very similar to Microsoft PowerPoint).
- Pixelmator (professional graphic editor, suitable for image editing).
- Corel PaintShop Pro (professional photo editor, suitable for editing images and photos).
- Paint.net (free image and photo editing software for windows computers).
- Gimp-Gnu (free image and photo editing software for most operating systems; has advanced features in comparison with Paint.net).
- LaTeX (good for layout, but requires other software applications, especially for drawing).
- Adobe Photoshop (suitable for image manipulation, but it is complex and expensive).
- Adobe Illustrator and InDesign (professional and produces higher resolution images).

This list is inexhaustive because it is possible that similar software will be developed and some in the list will be stopped after this book is published, and there might be plenty of similar excellent software packages out there that I am not aware of at the moment. Feel free to use any software as long it gives you the desired results – an effective and appealing scientific poster. Depending on the need of the author, the software packages can complement themselves (one for layout design, another for drawing, editing, *etc.*). Also, one software might be better than the other, depending on the author's need, *e.g.*, it is easier to create a large poster in Microsoft Publisher, Adobe InDesign, Scribus or QuarkXpress compared with Microsoft PowerPoint, which is suitable for creating relatively small posters. You might also need a software like ChemDraw to draw chemical structures. Because significant efforts go into layout design and editing of scientific posters, I strongly recommend that you get proficient at using any of the software packages available to you for poster making before you begin. YouTube is a great learning resource, especially with respect to software usage.

20.4   GETTING STARTED WITH THE POSTER

Before jumping to the computer to begin creating your poster for the conference or meeting, you need to be clear about certain things.

- *Abstract* and *storyline*: Agree the contents of the abstract with your mentor or supervisor and address all confidentialities and intellectual property issues before submitting it. The abstract should contain a single clear storyline (focusing on the main finding(s) of your research and should serve as the poster backbone.
- *Start early*: Once your abstract is acceptable for poster presentation, do not wait until the dying minute, start creating the poster well ahead of the conference or submission deadline (virtual session). This will allow you ample time to think carefully about the relevant stuff to include on the poster.
- *Software*: Decide the software you would like to create the poster with. You will probably need more than one software to create the poster, *i.e.*, a software for layout design, plotting and editing graphs, drawing sketches and illustrations. Once you have decided the software applications, learn to use them proficiently (the YouTube is a great learning resource) before using them to create your poster.
- *Sketch*: Decide the format (landscape or portrait) of the poster and make a rough sketch of its layout design on an A4 paper to serve as the skeletal framework.
- *Creation*: With all of the foregoing in place, you can go to the computer, set the desired poster dimension and begin creating the poster, bearing in mind who the audience are going to be (expert or nonexpert scientists). Expert scientists will be familiar with the fundamental concepts of the subject area and will understand the poster without conceptual explanation. Nonetheless, nonexpert scientists will not be familiar with the fundamental concepts of the subject area, making conceptual explanation necessary. Also, the poster dimension should allow a margin of at least 2 cm all around the poster.
- *Presentation*: Is the poster presentation going to be in-person or virtual? For in-person presentation, you will simply go to the venue and make your presentation. However, virtual presentation will require you to submit a pre-recorded presentation of the poster. Therefore, you will need an appropriate recording device and recording software. Such recording software are ubiquitous, including Microsoft PowerPoint, and the YouTube is a great place to learn how to use them.

20.5 COMPONENTS OF SCIENTIFIC POSTERS

The components of a poster vary with its purpose and the target audience. For scientific conferences or meetings, where the audience are mainly scientists who are familiar with the scientific style of reporting research findings, the posters are expected to contain chronologically:

- Title.
- Name of authors and their addresses/affiliations, with the presenting author clearly indicated.
- Abstract (optional) – Summary or brief overview of the research work.
- Background (replaceable with introduction) – Key information the reader needs to know so as to understand the work and the problem the research seeks to address or the research question.
- Methodology (materials and methods) – Key experimental procedures used in answering the research question.
- Results – Key experimental outcomes.
- Conclusion(s) – Key take-home message from the work.
- Acknowledgments – Expression of gratitude to those who have provided the money for the work or contributed to the success of the work.
- References (optional) – Relevant scientific papers, magazines, blogs, *etc.* consulted for the work.

These components and their contents (Gundogan et al. 2016) are much like those of a research paper (Chapters 3–13), but some of them have been subjected to debate. For example, the abstract and the reference sections are often considered optional or unnecessary (Lefor and Maeno 2016) components of scientific posters. I resonate with this for the following reasons:

- The abstract will surely be in the book of abstracts; therefore, there is no need to have it on the poster again.
- The abstract should not be part of the poster for the same reasons it is not part of oral presentation.
- Submitted abstracts are normally 250 words or so; therefore, having it on the poster will significantly contribute to the wordiness of the poster. Even if the abstract is further trimmed for the poster, that will lead to two different abstracts for a single research work.
- As an illustrative abstract, a carefully crafted poster will be readily understood compared with making sense from a 250 or so-word textual abstract.
- If the contents of the poster are already published, with the link given on the poster, then there is no need for the reference section as interested readers will find the complete list of references in the published work. However, if the work is not published, three key references will just be enough.

Therefore, scientific posters should necessarily contain: a *title*, list of *authors* and their *addresses*, the *background* or *introduction*, the *methodology*, the *results*, the take-home message (*conclusion*) and *acknowledgments*.

Having attended the poster session of many international conferences and meetings and judging posters at three different scientific meetings, I have observed three key problems with posters: *layout design* (*i.e.*, information placement order), *excessive information load* (*i.e.*, both as text and as graphics) and inappropriate *color* choice and overuse.

20.5.1 POSTER FORMAT AND LAYOUT DESIGN

There are two types of poster formats: *portrait* where the height of the poster is longer than its width and *landscape* where the height of the poster is shorter than its width. Both formats are acceptable at scientific conferences and meetings, but the organizers might stipulate their acceptable format in the *Call for Abstract* – be sure to use the appropriate format. Contrarily, three types of poster layout design can be identified in the literature, namely the story, the core (Rowe 2017) and the Mike Morrison layout designs.

a. *The story layout design*: This is the commonest poster layout design. The research finding is presented chronologically using the IMRaD (introduction, methods, results and discussion + conclusion) style of scientific research papers, with the various sections clearly indicated or implied (Rowe 2017), as illustrated in Figure 20.2 (my first scientific poster). The text in all the sections has the same font size and typeface. The main research findings (*i.e.*, conclusions in Figure 20.2) are also written in the same font size like the text in the other sections (*i.e.*, not emphasized), making it difficult for readers to readily grasp it.

b. *The core layout design*: Unlike the story layout design, the core layout design (Figure 20.3) emphasizes the main finding(s) by using a relatively larger text font size and placing it prominently at the center or upper part of the poster, thereby making it readily visible. In Figure 20.3, however, the key findings are not written in larger font, but they are highlighted and placed prominently in the upper part of the poster. With the main finding clearly visible, this layout design can attract readers' attention more readily than the story layout design and the message of the poster can also be readily gotten. The core layout design is more suitable for posters that do not require the IMRaD-style sections (Rowe 2017).

c. *The Mike Morrison layout design*: Similar to the core layout design, the Morrison layout design is "message-centered", with the main message boldly written in simple English and placed in the middle of the poster alone or with a graphic (*i.e.*, photo, graph or sketch), as shown in Figure 20.4. A quick response (QR) code (containing a link to the detailed study) is placed right below the main message for interested readers to access (Greenfieldboyce 2019) as described on YouTube (Morrison 2019). The Morrison layout design triggered the #betterposters campaign on Twitter, where examples of his layout design were showcased (Figure 20.4).

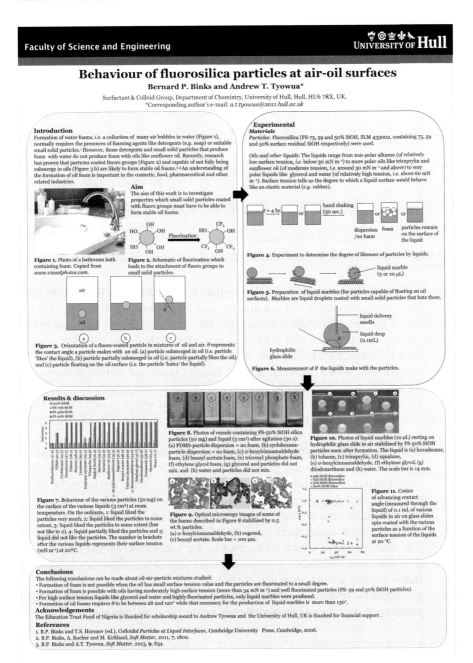

FIGURE 20.2 An example of a scientific poster prepared using the story layout design, showing the IMRaD structural style of scientific research papers (courtesy of Andrew Terhemen Tyowua).

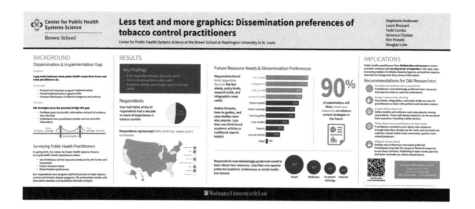

FIGURE 20.3 An example of a scientific poster prepared using the core layout design, showing the main findings prominently in the upper part of the poster under the result section (courtesy of Andersen et al. 2019).

Although the Morrison layout design is excellent because it has the IMRaD structure of scientific research papers and can also be used for posters that do not require this structure, it has few drawbacks. Firstly, the title is placed in the left-hand corner of the poster in a font size that is smaller than that of the main message, making the message look much like the title. Secondly, the design does not have room for acknowledgments, which is crucial as most funders insist that authors acknowledge them for funding their idea. Lastly, the sections are not numbered, so it might be difficult for readers to decipher exactly where to begin and where to stop. Therefore, I propose a modified version of the Morrison layout design in the following manner:

- The title should be written across the poster, with the various sections, including the acknowledgments, placed in boxes and numbered sequentially or linked by directional arrows to tell the reader where to begin and where to stop (naturally, from left to right and top to bottom). The boxes can be arranged in columns (landscape posters) or rows (portrait posters): a maximum of three columns for landscape posters and a maximum of four rows for portrait ones.
- The main message, which also serves as the conclusion, should be placed prominently at the center of the poster, but it should be written in a font size that is slightly smaller than that of the title.
- Finally, if the research work is published, a QR code containing a link to the publication page should be placed right below the main message.

This layout design is illustrated schematically in Figure 20.5 (landscape) and Figure 20.6 (portrait).

20.5.2 Excessive Information Load

The problem of excessive information load on a scientific poster can be overcome by mimicking the advertisement billboard (Figure 20.1). The billboard uses little text and appropriate photo, is straight to the point and is also readily understandable.

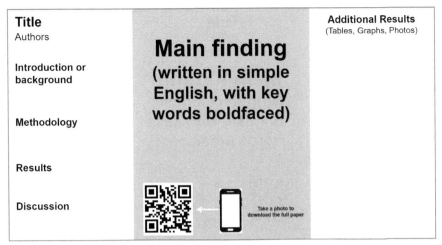

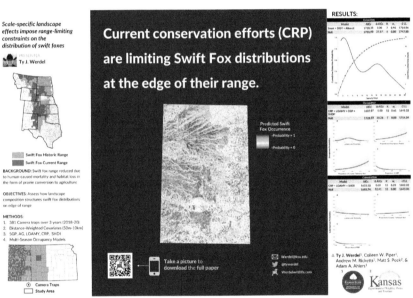

FIGURE 20.4 (a) Sketch of the Morrison layout design, showing the various sections along with a QR code. (b) An example of a poster prepared using the Morrison layout design (courtesy of Ty J. Werdel). The QR code contains a link to the publication page of the paper upon which the poster is based.

For each poster section, identify which information is important for the audience and which information is less important for them. Focus on the former rather than the latter, but aim for a good balance of empty space (30%), text (20–30%) and graphics (40–50%). Doing this spares the poster from being loaded with information that is not necessary for the audience, thereby making it much like the advertisement billboard. Such posters attract more audience because less time is required to read them (Rowe 2017).

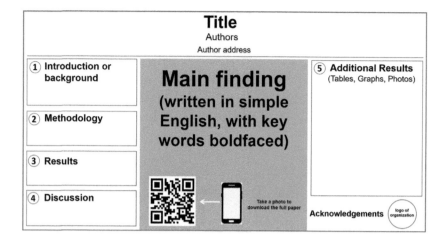

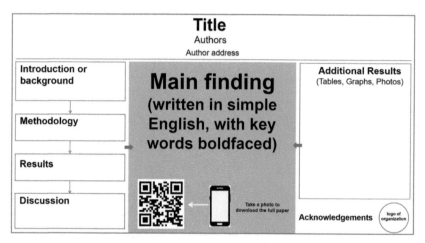

FIGURE 20.5 Sketches of the modified version of the Morrison layout design in landscape format, showing the various sections, including the acknowledgments section, numbered sequentially (upper) or linked by directional arrows. A QR code for accessing the published paper is also included.

20.5.3 INAPPROPRIATE COLOR CHOICE AND COLOR OVERUSE

Color perception is subjective, *i.e.*, one might be attracted by certain colors and be repelled by others, but the use of colors in poster making is not a mere matter of aesthetic subjectivity. Colors have meanings and implications (see Rowe 2017) which must be taken into account when preparing a scientific poster, irrespective of the author's aesthetic perception. The guiding rules in poster making that help check inappropriate color choice and color overuse are the following:

- Use of colors should be limited to graphics (graphs, charts, sketches, *etc.*).
- Chosen color themes should harmonize and unify the poster.

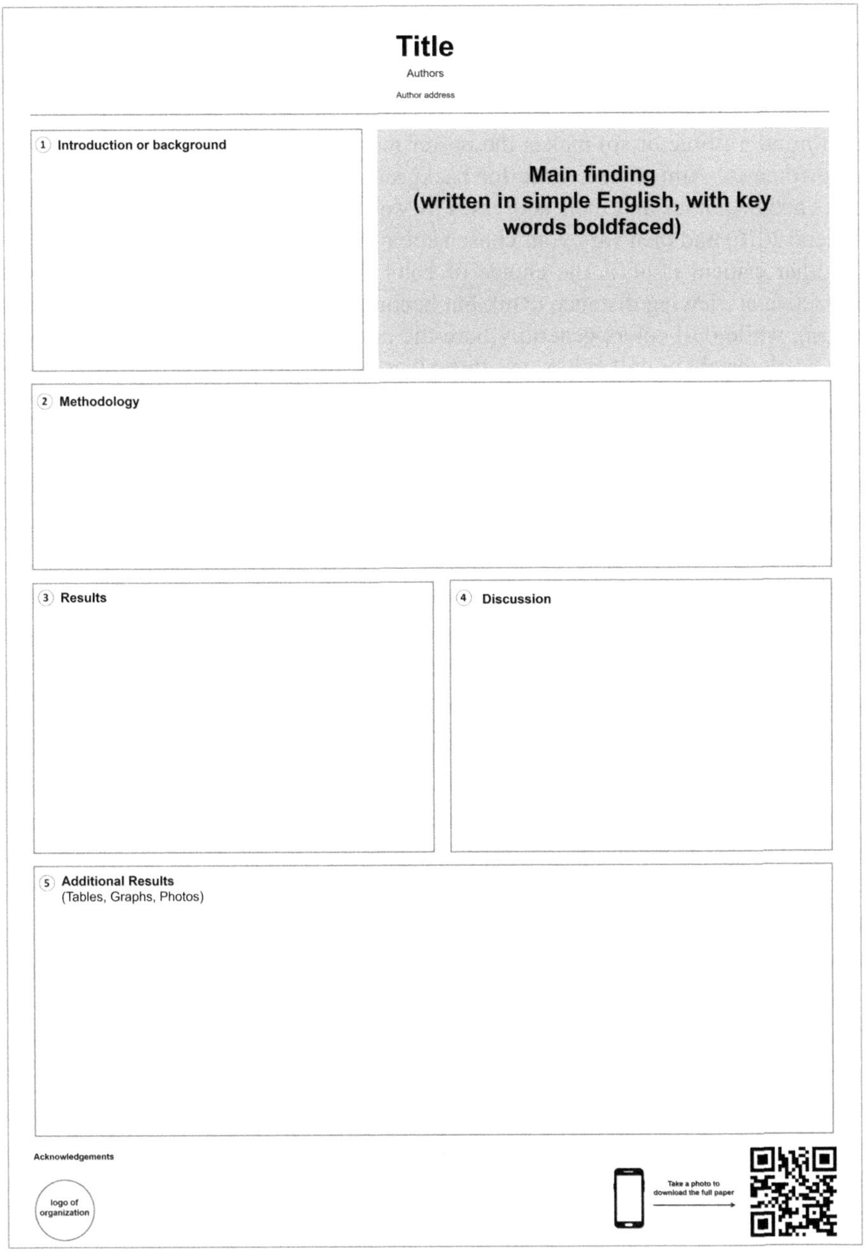

FIGURE 20.6 Sketch of the modified version of the Morrison layout design in portrait format, showing the various sections, including the acknowledgments section. A QR code for accessing the published paper is also included.

- Color themes should be used to differentiate related items on the poster.
- Color themes should support and project the main message of the poster.

Posters containing color themes that are based on these rules are generally more attractive than those with random, unharmonized color themes that differentiate nothing and add nothing to the message of the poster (Block 1996). Equally important is the number of color themes. Keeping the number of color themes to a minimum (three or so) makes the poster more attractive than those with several color themes. Aim to use white for background and black for text, including the title, author names and their addresses as well as the section headings (Lefor and Maeno 2016) and then vary your chosen color themes consistently across the poster. Another critical issue is the choice of color themes. Bright colors are generally attractive at viewing distance (3 m), but become overwhelming at reading distance (~2 m), while dull colors generally bore the poster (Rowe 2017). Therefore, do not use overly bright or dull colors; use those that are mid-way between these extremes. Such colors can be decided by carefully examining a color chart. Finally, be sure that text written on all color themes are readily readable.

20.5.4 POSTER SIZE, TITLE, SECTION HEADINGS, AUTHORS AND THEIR ADDRESSES AND BODY TEXT

a. *Poster size*: The conference or meeting organizing committee will prescribe the acceptable poster size and format in the *Call for Abstract*. For most scientific conferences and meetings, the poster size will be A0 while the format will be either landscape or portrait. Be sure to adhere to the recommended size to avoid disqualification from poster competition or even being debarred from participating at the event (Gundogan et al. 2016).

b. *Poster title*: A *Sans-serif* font type (*e.g.*, Helvetica, Arial, Calibra, Futura, Franklin Gothic, Tahoma, Verdana, *etc.*) is recommended for the poster title and section headings. Table 20.1 contains the font size range for text in the various sections of a scientific poster. Because readers are likely to be 1–3 m from the poster, and perhaps walking, the title and section headings should be respectively written in font size ≥ 80 and ≥ 30 points and bold-faced (Table 20.1).

TABLE 20.1

Ideal Font Size Range of Various Parts of a Scientific Poster

Component	Font Size/Point
Title (boldfaced)	80 – 100
Main message	60 – 70
Authors	30 – 60
Institutional address	20 – 40
Email and social media handle	20 – 35
Section headings (boldfaced)	30 – 50
Body text	24 – 40
Table and figure captions	20 – 35
Acknowledgments	20 – 28

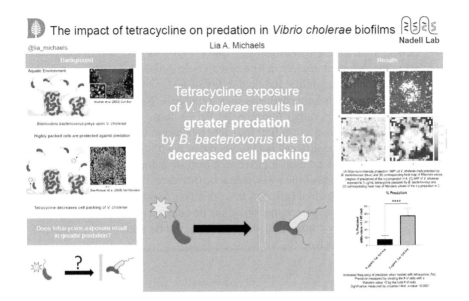

FIGURE 20.7   An example of a landscape scientific poster based on the Morrison layout design with the title written across. (Courtesy of Lia Michaels.)

In fact, the title text should be the biggest on the poster. Use sentence case rather than capital letters, which are difficult to read (Block 1996, Kojima and Patrick Barron 2016), for both the title and section headings. The title should be ten words or so, and it should summarize the work in a concise and inviting manner. The title should be placed across the poster, as shown in Figure 20.7. This style ensures that the title is clearly visible and psychologically tells the reader that everything underneath is supporting the poster title. The title may be either descriptive, in the form of a question or as a statement announcing the experimental finding. The title should be informative to both experts and nonexperts (Erren and Bourne 2007), especially if it is not a specialist event, and it should not contain acronyms, abbreviations and subject specific jargons.

c. *Authors' names and their addresses*: Author names (font size ≥ 30 points, Table 20.1) should be placed just below the poster title while their addresses/affiliations (font size ≥ 20 points) can be placed just below their names. For multi-authored posters, the email addresses and social media handles (Twitter, Facebook, LinkedIn, *etc.*) of the presenting author can also be included (font size ≥ 20 points), especially if the author shares his/her research work on them. The author names and their addresses can also be written in Sans-serif font type using sentence case, but they should neither be boldfaced nor italicized. Author names can either be written in full (*e.g.*, Bernard Paul Binks) or initialized (*e.g.*, BP Binks), especially for a long author list. If there are more than five authors, the name(s) of the principal author(s) can be supplied on the poster while a link for the rest can be created and supplied on the poster.

d. *Body text*: The Serif font type like Times New Roman or its clones like Georgia, Garamond, Courier new, Century, Baskerville, Palatino Schoolbook, *etc.* is recommended (Block 1996) for text within the poster sections. Although widely debated (see Poole 2008), this recommendation is based on the legibility and readability of text written in these typefaces compared with those written in others typefaces. Nonetheless, my recommendation here is rather because the Serif's font type, especially Times New Roman, is more familiar and comprehensible to readers as it is commonly used in scientific publications compared with other typefaces. Stylish typefaces that are not familiar to readers will slow down their reading speed as they try to decipher the text characters; therefore, they should be avoided. Finally, the chosen font type and size (ideally ≥ 24 points) should be consistent throughout the poster and should not be boldfaced, except when emphasizing certain keywords. Also, the text should not be justified or centralized, but should align left.

20.5.5   BACKGROUND, OBJECTIVE, METHODOLOGY AND RESULTS SECTIONS

a. *Background*: In the background or the introductory section, use two to four bullet points or graphics (photo, sketches, graph, *etc.*) or a combination of bullet points and graphics to tell the audience what they need to know to understand your poster. This should include the motivation for the work and the problem you have set-out to solve or the question you have set-out to answer.

b. *Objective*: State the purpose of your experiments in a sentence or a graphic, *e.g.*, to solve the problem or answer the question posed in the background section. For a poster, the objective can also be considered as the aim of the work.

c. *Methodology*: Use sketches, illustrations, flowcharts, *etc.*, with little text (for explanation) where necessary, to describe your experiments. Limit your description to key experiments and link-up experimental steps with directional arrows. You can use short video clips to support your methodology. This can be achieved by uploading the said video online and inserting the link or the QR code on the poster. You can also carry the video on your tablet and play it for your audience when the need arises.

d. *Results*: Present your results in the form of linear graphs, bar charts, histograms, photos, micrographs, *etc.* Experimental data should be presented as a linear graph, bar chart, *etc.* instead of tables. Photos, plotted graphs and charts, especially those from Microsoft Excel, should be edited appropriately (see Chapter 11) before use. These must be of high resolution (≥ 300 dot per inch). However, present the key result(s) and move the rest to the additional figures and tables section of the poster. The caption for each result should be brief and should highlight the finding. Just like the methodology section, you can also use short video clips to support the results section. This can be achieved *via* the same ways described in the methodology section.

The appropriate verb tenses for these sections are the same with those of scientific research papers summarized in Chapter 14.

20.5.6 Conclusion Section and QR Code

a. *Conclusion*: The key conclusion derived from the experiment should be boldly written (font size ≥ 60 points, Table 20.1) as the take-home message. This should be written in simple English in the middle of the poster, with the keywords boldfaced.

b. *QR code*: The QR code is a matrix barcode invented by Denso Wave Company, Japan, in 1994. Information about an item is stored on the barcode and placed on the item as an optical label. This information is retrieved by scanning the barcode using an appropriate device, *e.g.*, a smartphone with an appropriate barcode scanner application. For a scientific poster, a link to the published paper or details of the work can be saved on the barcode and placed on the poster for the audience to access. Barcodes can be generated free of charge at https://www.qr-code-generator.com/ or other similar websites.

An example of a landscape scientific poster (size A0), prepared from Tyowua et al. (2022), in line with the suggestions given here, is shown in Figure 20.8. The poster is prepared using a combination of three software, namely Microsoft Publisher (layout design), Microsoft PowerPoint (sketches/illustrations) and Microsoft Excel (plotting graph). The font type and size used are summarized in Table 20.2.

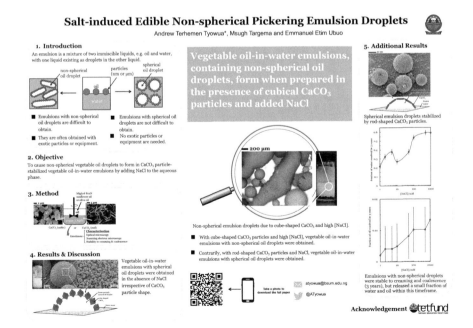

FIGURE 20.8 An example of a scientific poster (landscape) prepared in line with the suggestions given here for creating posters. (Courtesy of Andrew Terhemen Tyowua.)

TABLE 20.2

Font Size of Various Parts of the Scientific Poster shown in Figure 20.8, Including the Line and Marker Thickness Used on the Graph

Part	Font Size or Line Thickness/Point
Title (boldfaced)	80
Main message (boldfaced)	70
Authors	45
Email and social media handle	30
Section headings (boldfaced)	45
Body text	35
Table and figure captions	32
Graph	
Axis title	18
Axis values	18
Border lines	2.25
Trend lines	2
Data marker	10

The Georgia font type is used throughout the poster except for the author names, email address and social media handle where the Arial font type is used.

Activity 20.1 *Reading Scientific Posters*

Read the scientific posters in Figures 20.2–20.4, 20.7, and 20.8 as though there were displayed at a scientific conference. Which of these posters did you understand readily? For those that you did not understood readily, what features do they have in common?

20.6 POSTER PRESENTATION

At the scientific conference or meeting, you will be assigned a space to hang your poster where readers will come around, typically during tea breaks or poster sessions, to read your work, ask questions and interact with you. Readers who are really interested in the work might spend up to 30 min or more discussing and asking questions about the work. However, research has shown that on average, readers spend 3–5 min at a poster (Block 1996). Therefore, it is advisable to read your poster through once it is completed, noting how long it takes to read it through and how readily it can be understood. You can equally ask your friends or colleagues to read the poster and, similarly, give you feedback. Edit and improve the poster if it requires more time (> 5 min) to read and understand until it meets this important criterion before printing (in-person session) or uploading (virtual session) it. At the conference or meeting, you are expected to stand by your poster to answer questions and interact with the audience. Take this as a great opportunity to discuss your work and get direct feedback about it and also develop professional networks and contacts. This entails having a clear plan for the presentation:

- Rehearse a 1 to 4-minute explanation of your poster, with a focus on the big picture (*i.e.*, main message), but allow readers to first read on their own before asking if they would like you to talk them through the poster. In your rehearsal, pay particular attention to the graphics and sketches and be ready with a clear explanation for them. Also, be ready to discuss future direction of the work as this will likely come up at some point in your presentation.
- Make copies of the scale down version (*e.g.*, A4) of the poster for interested readers that might want a copy. Also, be ready with your business card to hand out to readers that might want to keep in touch with you. Finally, keep a notepad by to make notes and draw illustrations during the presentation.
- Wear an attire that coordinates with the color themes of the poster as this will likely draw more audience (Keegan and Bannister 2003).
- The poster should aid your explanation, but do not read it to the audience.
- Your discussion should be limited to the message of the poster. Do not delve into issues your work did not address.
- Admit your inability to answer questions outside the scope of your study or make an educated guess or speculation. You can also seek the questioner's thoughts.
- Admit your inability to answer questions that are within the scope of your research but are rather difficult for you. In this case you can refer the questioner to where he/she will likely find the answer or promise to find out the answer and get back to him/her.
- If an audience persistently asks about what you do not have answers for, suggest a private and more engaging discussion after the session.
- Keep calm with hostile audience who might ask ridiculous questions about your research and respond with a professional statement about the subject area or a related subject area.
- Overall, be calm and polite in all your responses.
- Finally, if possible, find out about the research interest of the audience for a possible chance to collaborate.

With travel restrictions and the encouragement of social distancing due to the Covid-19 pandemic, poster presentation can also be virtual. A virtual poster presentation usually involves:

- Uploading the poster (as a photo) to an online gallery (with or without audience discussion boards), where the audience view and discuss them either offline or online (if there are audience discussion boards).
- Uploading a pre-recorded video or audio presentation of the poster (video making and editing software are absolutely crucial here, see Chapter 5 for tips), where the content of the poster is presented in say 3–5 min, with the audience asking questions thereafter. For example, watch "Salt-induced edible non-spherical Pickering emulsion droplets" on YouTube.
- Making a live video presentation of the poster (3–5 min) using an online meeting platform, with follow up questions or interaction with the audience.

To successfully upload a photo of the poster or a video, be sure to adhere to the size and format stipulated by the organizers. If necessary, make use of photo and video editing software to convert between file formats and also reduce file size.

Activity 20.2 *Creating and Presenting a Scientific Poster*

Using the suggestions given in this chapter, create a scientific poster from any research work of your choice and make a 4-minute video presentation of the poster.

FURTHER READING

Carter, M. 2012. *Designing Science Presentations: A Visual Guide to Figures, Papers, Slides, Posters, and More*: Academic Press, London.
- Contains a detailed discussion of the visual elements (*i.e.*, graphs, sketches, photographs and text) of scientific presentations and gives practical suggestions and guide for preparing scientific posters.

Nicol, A.A., and P.M. Pexman. 2003. *Displaying Your Findings: A Practical Guide for Creating Figures, Posters, and Presentations*: American Psychological Association, Washington.
- Contains useful suggestions for creating scientific posters and oral presentations.

Faulkes, Z. 2021. *Better Posters: Plan, Design and Present an Academic Poster*: Pelagic Publishing Ltd, Exeter.
- A specialized textbook on poster making. It contains detailed practical suggestions for creating and presenting scientific posters.

Rougier, N.P., M. Droettboom, and P.E. Bourne. 2014. Ten simple rules for better figures. *PLoS Comput. Biol.* 10 (9):e1003833.
- Contains ten suggestions for creating better figures for scientific research papers, posters and oral presentations.

Rowe, N. 2017. *Academic & Scientific Poster Presentation*: Springer, Switzerland.
- Contains a detailed overview of academic and scientific poster preparation and presentation.

REFERENCES

Akister, J., A. Bannon, and H. Mullender-Lock. 2000. Poster presentations in social work education assessment: a case study. *Innov. Educ. Train. Int.* 37 (3):229–233.

Andersen, S., L. Brossart, T. Combs, V. Chaitan, P. Kim, and D. Luke. "Less text and more graphics: Dissemination preferences of tobacco control practitioners." Poster presentation at the 12th Annual Conference on the Science of D&I in Health, December 2019. Center for Public Health Systems Science at the Brown School at Washington University in St. Louis.

Block, S.M. 1996. Do's and don't's of poster presentation. *Biophys. J.* 71 (6):3527–3529.

Bracher, L.E.E. 1998. The process of poster presentation: A valuable learning experience. *Med. Teach.* 20 (6):552–557.

Chopra, R., and A. Kakar. 2014. The art and science of poster presentation in a conference. *Curr. Med. Res. Pract.* 4 (6):298–304.

Costa, M.J. 2001. Using the separation of poster handouts into sections to develop student skills. *Biochem. Mol. Biol. Educ.* 29 (3):98–100.

Erren, T.C., and P.E. Bourne. 2007. Ten simple rules for a good poster presentation. *PLoS Comput. Biol.* 3 (5):e102.

Greenfieldboyce, N. 2019. "To save the science poster, researchers want to kill it and start over." Npr. https://www.npr.org/sections/health-shots/2019/06/11/729314248/to-save-the-science-poster-researchers-want-to-kill-it-and-start-over?utm_campaign=storyshare&utm_source=twitter.com&utm_medium=social.

Gundogan, B., K. Koshy, L. Kurar, and K. Whitehurst. 2016. How to make an academic poster. *Ann. Med. Surg.* 11:69–71.

Hess, G.R., and E.N. Brooks. 1998. The class poster conference as a teaching tool. *J. Nat. Resour. Life Sci. Educ.* 27 (1):155–158.

Hess, G.R., K.W. Tosney, and L.H. Liegel. 2009. Creating effective poster presentations: AMEE Guide no. 40. *Medical teacher* 31 (4):319–321.

Keegan, D.A., and S.L. Bannister. 2003. Effect of colour coordination of attire with poster presentation on poster popularity. *Can. Med. Assoc. J.* 169 (12):1291–1292.

Kojima, T., and J. Patrick Barron. 2016. Making the most of your poster presentation (1): Poster design. *Jpn. J. Gastroenterol. Surg.* 49 (1):72–73.

Lefor, A.K., and M. Maeno. 2016. Preparing scientific papers, posters, and slides. *J. Surg. Educ.* 73 (2):286–290.

Mitrany, D. 2005. Creating effective poster presentations: the editor's role. *Science* 28 (4):114–116.

Moneyham, L., D. Ura, S. Ellwood, and B. Bruno. 1996. The poster presentation as an educational tool. *Nurse Educ.* 21 (4):45–47.

Morrison, M. 2019. How to create a better research poster in less time (#betterposter Generation 1). YouTube: https://www.youtube.com/watch?v=1RwJbhkCA58.

Moule, P., M. Judd, and E. Girot. 1998. The poster presentation: What value to the teaching and assessment of research in pre- and post-registration nursing courses? *Nurse Educ. Today* 18 (3):237–242.

Nundy, S., A. Kakar, and Z.A. Bhutta. 2022. *How to Practice Academic Medicine and Publish from Developing Countries? A Practical Guide*: Springer Nature, Singapore.

Pelletier, D. 1993. The focused use of posters for graduate education in the complex technological nursing environment. *Nurse Educ. Today* 13 (5):382–388.

Poole, A. 2008. "Which are more legible: Serif or Sans Serif Typefaces?". http://alexpoole.info/blog/which-are-more-legible-serif-or-sans-serif-typefaces/.

Rowe, N. 2017. *Academic & Scientific Poster Presentation*: Springer, Switzerland.

Sousa, B.J., and A.M. Clark. 2019. Six insights to make better academic conference posters. *Int. J. Qual. Methods* 18:1–4.

Tyowua, A.T., M. Targema, and E.E. Ubuo. 2022. Salt-induced edible anisotropic Pickering emulsion droplets. *J. Dispers. Sci. Technol.*:1–12.

.

21 Oral Presentations

21.1 ORAL PRESENTATION IN SCIENCE

An oral presentation is a verbal communication where the presenter speaks to a group of audience using printed text or projected text from slides. Oral presentations are often used to disseminate scientific findings at conferences and meetings. The presenter projects prepared slides onto large screens during the presentation and the audience ask questions and/or make contributions after the presentation. An oral presentation can be thought of as a scale-up poster presentation because the former is slightly more elaborate than the latter and, to a large extent, the principles for preparing them are also the same. Similar to posters, there are plenty of scientific oral presentations that are simply below the mark in that they are boring and difficult to follow. Therefore, this chapter gives practical suggestions for preparing and giving a better oral presentation at scientific conferences and meetings. With minor modifications, these suggestions can also be used to prepare and give oral presentations for other purposes, *e.g.*, student oral assessment, *viva voce*, lecture, *etc.*

21.2 FEATURES OF A GOOD SCIENTIFIC ORAL PRESENTATION

A good scientific oral presentation (Estrada et al., 2005):

- Contains a clear single central message.
- Contains the right order of information, preferably the IMRaD structure of scientific research papers.
- Engages the audience (*i.e.*, excites interest by showing the importance of the findings).
- Contains the right amount of text (20–30%), graphics (*i.e.*, graphs, sketches and photos, 40–50%) and empty space (30%).
- Contains a well-blended color mix of graphics.
- Contains the right font type and text size.
- The presenter (i) speaks to the microphone (*i.e.*, is clear and audible) and uses the laser pointer only when it is necessary; (ii) maintains an appropriate or natural speech pace (*i.e.*, not too fast nor too slow) (Busà, 2010); (iii) shows significant authority of the subject area, *i.e.*, knows the topic inside out; (iv) maintains a good eye contact with the audience; and (v) gives clear, appropriate and succinct responses to questions.
- Does not exceed the allotted time.

DOI: 10.1201/9781003186748-21

21.3   SOFTWARE PACKAGES FOR PREPARING SCIENTIFIC ORAL PRESENTATIONS

The following software packages can be used to prepare slides for oral presenta-tion: Microsoft PowerPoint, Google slides, Prezi, Slidebean, Slide Dog, Keynote, LibreOffice Impress, Zoho show, WPS Office or Visme Presentation tool. This list is inexhaustive because it is possible that similar software packages will be developed or some of those on the list will be stopped after this book is published. Therefore, feel comfortable to prepare your slides with any software package available to you. However, to be on a safer side, check with the organizers that such slides will be compatible with their computers. If you need help with using a software, remember that the YouTube is a great place to learn how to use any software for free.

21.4   PARTS OF A SCIENTIFIC ORAL PRESENTATION

A typical scientific oral presentation contains the IMRaD structure of scientific research papers (Vučković-Dekić, 2002, Wellstead et al., 2017). However, the con-tents of the sections are not as elaborate as those of a scientific research paper.

- *Introduction (or background)*: Essential information about the work to bring the audience to speed, including the importance of the work (Horiuchi et al., 2022) and a statement of the research problem or question.
- *Methodology*: Description of materials, equipment and experimental pro-cedure used, but not to enable reproducibility, *i.e.*, not as detailed as that in a research paper (Vučković-Dekić, 2002). The methodology section is not part of some oral presentations.
- *Results*: Presentation of experimental data or outcome of experimental investigation. This section can be combined with the discussion section.
- *Discussion*: Interpretation and explanation of experimental data or outcome. The results and the discussion sections are the most important parts of oral presentations and take ~80% of the presentation (Vučković-Dekić, 2002).
- *Conclusion(s)*: Summary of the take-home information from the research.
- *References*: A long list of references is not required, but rather a few rel-evant references (scientific papers, magazines, blogs, *etc.*), especially semi-nal papers, can be mentioned during the presentation.
- *Acknowledgments*: Expression of gratitude to the funder and those that have helped in one way or the other.

21.5   GETTING STARTED WITH ORAL PRESENTATION

The success of an oral presentation depends strongly on four aspects: (i) planning, (ii) creating, (iii) practicing and (iv) presenting.

21.5.1   PLANNING

The old adage which says "he who fails to plan, plans to fail" also applies to oral presentation. A good oral presentation cannot be achieved without careful planning. An ideal oral presentation plan involves:

- *Agreeing the contents*: It is difficult to include all the data, even for long talks; therefore, decide (Vučković-Dekić, 2002) or agree the contents of the presentation with your mentor or supervisor. The agreed contents should lead to a clear single message (Katchburian, 2003). Alternatively, identify the main finding of the research, with the supporting results, and form a clear single message for the presentation. Either way, the contents of the presentation should be decided bearing in mind all confidentialities and intellectual property issues. Once the contents and the message have been decided, make a resonating 25- to 30 word summary of the message, preferably with everyday words, to serve as the framework of the presentation.
- *The audience*: Knowing who the audience are, experts or nonexperts scientists, determines the language and contents of the oral presentation (Vučković-Dekić, 2002, El Sabbagh and Killu Ammar, 2015). Experts will be familiar with the jargons of the area, while nonexperts may not understand the jargons and the contents; therefore, the language must be reframed to enable them to understand.
- *Purpose*: It is important to know the primary purpose of your presentation before you start putting the slides together. For example, the purpose of an oral presentation at a scientific conference or meeting is to highlight findings of a research work in 8–12 minutes, stimulating interested audience to read the published paper. This is different for an interview presentation, where the aim is to showcase your research focus to a broad spectrum of audience within and outside your subject area as well as demonstrate your ability to transfer knowledge so as to get hired. A conference talk is also different from a class presentation where the main aim is to impress your professor with your knowledge of the subject and earn high grade. All of these presentations can be approached differently, with the suggestions given here for conference talks as the reference point. Generic suggestions about oral presentations can be found in Storz (2002).
- *Format and slide size*: The IMRaD structure of scientific research papers is ideal for oral presentation and should be used because it is the acceptable scientific reporting style. Although slight variations are acceptable, it is advisable to adhere to the IMRaD structure. If you are preparing your slides in Microsoft PowerPoint, remember that you can choose between the standard slide size (4:3) or the widescreen size (16:9). You can also customize the slide size altogether. Nonetheless, whatever software package you are using, be sure that when projected, the slides are going to fit well onto the projector screen. Although, in some cases, fitting of slides onto the projector screen is achieved through the projector settings. The conference organizers can help you decide the appropriate slide size to use.
- *Timing*: For most conferences and meetings, the time allotted for an oral presentation is 8–12 minutes. Plan to finish your presentation within this timeframe so that you are not stopped in the middle of it (Katchburian, 2003, Bulska, 2006). That means going by the ideal benchmark of not more than two slides per minute, you should plan to have around 19 slides, excluding the opening and the outline slides, for an 8-minute presentation.

Nevertheless, bear in mind that some slides (*e.g.*, those with graph) will require more time to explain while other slides (*e.g.*, those with photos or only text) will require less time to talk through.

21.5.2 Creating

Projectors and projector screens are often available at many scientific conferences and meetings to aid oral presentations. During presentation, the presenter is expected to project their points onto the screen for the audience to follow along. Therefore, it is imperative to create visually appealing and informative slides for the audience. You can create the slides using any software package available to you, but they must be compatible with the gadgets available for displaying them. The following suggestions will help you prepare visually appealing and informative slides:

i. Endeavor to use bullet points throughout and aim to have four to seven bullet points per slide, with each point containing four to seven words. Short phrases are especially important here (Estrada et al., 2005, Papanas et al., 2011, Wellstead et al., 2017). Cowan (2001) showed that a human being remembers about four pieces of information if presented with multiple pieces of information in a short space of time. Therefore, Horiuchi et al. (2022) recommend that major concepts should be limited to four points and that four points per slide is more ideal.
ii. Aim to use more graphics (sketches, illustrations, graphs, *etc.*) and less text and try to alternate between slides with only text and those with only graphics (Wellstead et al., 2017).
iii. Your tables and graphs should have thick lines so that they will be visible to all the audience in the hall (if you are in doubt, see Rougier et al., 2014).
iv. Your tables and graphs should be accompanied by captions that state the main point they contain (Horiuchi et al., 2022).
v. Use a Sans-serif font typeface (Estrada et al., 2005) and a relatively large font size ≤ 36 points (El Sabbagh and Killu Ammar, 2015), *e.g.*, 32 points (boldfaced, section headings) and font size 24–28 points (not boldfaced, general text within sections).
vi. Formatting (margins on all sides), font type and size should be consistent throughout the slides (Horiuchi et al., 2022).
vii. Use colors sparingly and stick to chosen color themes throughout the slides (Horiuchi et al., 2022).
viii. Boldface or italicize words to emphasize them but do not underline or capitalize them as this will be difficult to read for audience far from the projector screen.
ix. Aim to use a white background with dark text (small audience), dark background with white text (large audience) or light-colored background with dark text (Wellstead et al., 2017).
x. Proofread and correct all spellings and grammatical errors.

These suggestions can be summarized into two acronyms: KISS (Keep It Short and Simple) and KILL (Keep It Large and Legible) (Kanpolat, 2002). The slides can be prepared in the following tentative chronological order:

- *Opening*: The opening slide should contain the (i) title or topic of the presentation (see types of titles in Chapter 3); the title should be clear, succinct and informative; (ii) author name(s); (iii) author address/affiliation; and (iv) date of the presentation.
- *Conflict of interest (optional)*: Some conferences now require presenters to declare any form of conflict of interest (Alexandrov and Hennerici, 2013); therefore, this slide should contain a list of all forms of conflict of interest in connection with the work presented.
- *Forecast (optional)*: This slide should contain a brief statement of the problem tackled and what was found; this slide serves as the "abstract" of the talk.
- *Outline*: This slide should give the structure of the presentation in the form of a table of contents or an outline.
- *Introduction (or background)*: This slide should contain (i) essential information necessary for understanding the talk, (ii) motivation and importance of the work (why should the audience care about the work?), (iii) statement of the problem or research question, and (iv) a summary of key related researches (if published, refer audience to the paper for full details).
- *Methodology*: This slide should contain a brief description of the materials and experimental procedures (if published, refer the audience to the paper for details).
- *Results and discussion*: This section forms the main body of the presentation; therefore, it should be properly crafted and should be given priority during the presentation. This section should contain key results that support the main theme of the talk and sufficient interpretation/explanation of the results. This section should also contain the significance of the result, *i.e.*, why should the audience care about the findings? Numerical data should be presented as nice graphs or charts (see Rougier et al., 2014 for guide) rather than in a tabular form. Each slide in this section should contain a single idea.
- *Conclusion/summary*: This slide should contain a resounding take-home message of the research work; use a summary diagram, if possible.
- *Future work (optional)*: This slide should contain questions that the research work has opened and you plan to investigate in future.
- *Acknowledgments*: This slide should contain a list of funders and those that have contributed to the overall success of the work.
- *Appreciation (optional)*: This slide simply says "thank you for listening", but other related phrases can also be used.
- *Appendix (optional)*: This slide is not part of the main talk, but it is useful for answering questions and clarifying issues. Slides in this section should contain additional results on potential question areas, supporting evidence on questionable experimental procedures, negative results and derivation of mathematical formulae.

The appropriate verb tenses for these sections are the same with those of scientific research papers summarized in Chapter 14.

21.5.3   PRACTICING

It is important to practice your presentation verbally (not mentally) once you have finished preparing it (Katchburian, 2003) and also read papers on the topic to improve or cement your knowledge of the topic. Practice will increase your confidence and help you overcome stage nervousness (Roberts, 2013), while reading related papers will help you with the necessary responses to questions. Practice before friends or colleagues who will criticize and give you useful feedback (Alexandrov and Hennerici, 2013, Wellstead et al., 2017). You can also practice alone if you have no one around. When practicing alone, make a video of your presentation and review it thereafter, noting the time taken, talking speed, your gestures and pauses. You might notice that you are exceeding the time, doing some distracting gestures, having awkward pauses, talking fast or slow. Practice repeatedly, making videos and reviewing them with the aim of fixing all of these issues before the D-day. You will surely become better, more comfortable and confident, the more you practice.

21.5.4   PRESENTING

Many people are nervous about giving a talk, especially before experts who have years of experience on the subject area for fear of criticism and scrutiny. This makes them either noticeably shy or causes them to talk too fast; thus, impacting the quality of their presentation. The fact is that no one, no matter their years of experience, will know your work better than you (Roberts, 2013). Also, the big names in the subject area, who have come to listen to you, have done so because they want to learn from your work rather than to scrutinize and criticize it. The success of your presentation also depends on your ability to prevent technical problems that might impact it, the way you give it and the way you answer follow-up questions. These can be mitigated by being proactive before the presentation and avoiding or doing certain things during the presentation as well as during the follow-up question session.

 a. *Before presentation*
    Arrive early (at least 30 minutes) before the session commences. Within this time,
    • hand in the electronic copy of your presentation to the information technology (IT) staff,
    • check and address operating system compatibility issues with the IT staff,
    • test run the sound system and any videos in your presentation (Katchburian, 2003, El Sabbagh and Killu Ammar, 2015). Have a backup plan in case your video does not work during the presentation (Wellstead et al., 2017) and
    • familiarize yourself with the podium and how to advance slides (with the mouse, the keyboard or laser pointer) (Alexandrov and Hennerici, 2013).
 b. *During presentation*
    Appear appropriately dressed and smart, if possible. In many cases, the opening slide will be displayed, the moderator will introduce you (and the research team) and announce the title of your talk. If this is the case, thank

the moderator for introducing you and begin your presentation (Alexandrov and Hennerici, 2013). Contrarily, if the opening slide was not displayed, you were not introduced and the title of your talk was not announced, display the opening slide once you reach the podium and begin the presentation with a self-introduction and announce the title of the talk. During the presentation,

- make the audience comfortable and engaged with the slides and yourself by exciting interest in the topic (the introductory slide is very useful for this);
- maintain eye contact with the audience as you present, *i.e.*, look at the slides briefly and gaze more at the audience (Katchburian, 2003), pausing briefly on some of them and giving them a smile, when necessary;
- maintain an appropriate speech speed, tone and body posture and move about the space (especially for a small audience), if possible;
- speak to the audience and do not read the slides verbatim (Vučković-Dekić, 2002, Katchburian, 2003, Alexandrov and Hennerici, 2013);
- use gestures to stress important points, pronounce words clearly and use natural intonation with stress on important words and pause briefly after key points (Papanas et al., 2011). Clear pronunciation of words makes you understandable by both native and non-native speakers of English (Alexandrov and Hennerici, 2013);
- use the laser point sparingly, *i.e.*, to point at important features on the slide rather than pointing at everything on the slide (Papanas et al., 2011, Wellstead et al., 2017);
- when you display any graphics, tell the audience what it is (Alexandrov and Hennerici, 2013) (*e.g.*, plot of concentration against rate) and state the information it contains (*e.g.*, rate increases as concentration increases);
- minimize the use of filler words like "I mean, okay, em", *etc.*, but rather use rhetorical questions occasionally to maintain audience attention (Wellstead et al., 2017);
- do not turn your back on the audience to look at the slides (Bulska, 2006, Alexandrov and Hennerici, 2013) or cover the screen with your body;
- slow down near the end and convey the take-home message as clearly as possible. Using phrases like "in summary", "in conclusion", "to summarize", *etc.*, often help call back the attention of audience that has drifted. Also, leaving the conclusion/summary slide for a few minutes helps the audience to digest the message better (Wellstead et al., 2017).

c. *Question and answer session*

Most oral presentations are accompanied by a question and answer session where the audience ask the presenter questions for clarification or make their contributions (Vučković-Dekić, 2002). You should expect three question types: a genuine question for knowledge or clarification, a selfish question aimed at drawing attention to the questioner's ingenuity or a malicious question aimed at running you down. The following steps will help you to successfully sail through this session (Vučković-Dekić, 2002), irrespective of the question type:

- Listen carefully and write down the question, if possible, or ask someone in the audience to help. Thank the questioner and acknowledge the good impact of the question.
- Think carefully before answering; during this time, rephrase the question in your own words to be sure that you understand it and the audience have heard it clearly too.
- For a genuine question, answer it clearly and succinctly, if you can, and check that the questioner is satisfied with your response by asking if your answer makes sense to them or is clear. Limit your response to the question (Estrada et al., 2005).
- For a selfish or a malicious question, which you might not be sure of the answer, thank the questioner for raising an important question, but say you need to look further for a more robust response and then offer to discuss it one-on-one after the talk.

Overall, your responses must be polite and professional regardless of the question type, and it is alright to say "I do not know" when you do not have an answer to a question (Vučković-Dekić, 2002).

FURTHER READING

Carter, M. 2012. *Designing science presentations: A visual guide to figures, papers, slides, posters, and more*: Academic Press, London.
- Contains useful and practical suggestions for creating and giving oral presentations.
Gallo, C. 2010. *Presentation secrets of Steve Jobs: How to be insanely great in front of any audience*. McGraw-Hill Education, New York.
- Describes the techniques Steve Jobs, one of the world's best communicator, uses to give mind-blowing presentations.
Horiuchi, S., J.S. Nasser, and K.C. Chung. 2022. The art of a scientific presentation: Tips from Steve Jobs. *Plast. Reconstr. Surg.* 149 (3):533–540.
- Contains useful tips, derived from Steve Jobs presentation style, for giving effective scientific presentations.
Nicol, A.A., and P.M. Pexman. 2003. *Displaying your findings: A practical guide for creating figures, posters, and presentations*. American Psychological Association.
- Contains useful and practical suggestions for creating and giving an oral presentation.
Rougier, N.P., M. Droettboom, and P.E. Bourne. 2014. Ten simple rules for better figures. *PLoS Comput. Biol.* 10 (9):e1003833.
- Contains ten suggestions for creating better figures for scientific research papers, posters and oral presentations.
Stapleton, P., A. Youdeowei, J. Mukanyange, and H. Van Houten. 1995. *Scientific writing for agricultural research scientists*. Vol. 20, Ede: WARDA/CTA.
- Contains suggestions for preparing and giving oral presentations, albeit for agricultural scientists, but chemical scientists can also find them useful.

REFERENCES

Alexandrov, A.V., and M.G. Hennerici. 2013. How to prepare and deliver a scientific presentation. *Cerebrovasc. Dis.* 35 (3):202–208.

Bulska, E. 2006. Good oral presentation of scientific work. *Anal. Bioanal. Chem.* 385 (3):403–405.

Busà, M.G. 2010. Sounding natural: Improving oral presentation skills. *Language Value* 2:51–67.

Cowan, N. 2001. The magical number 4 in short-term memory: A reconsideration of mental storage capacity. *Behav. Brain Sci.* 24 (1):87–114.

El Sabbagh, A., and M. Killu Ammar. 2015. The art of presentation. *J. Am. Coll. Cardiol.* 65 (13):1373–1376.

Estrada, C.A., S.R. Patel, G. Talente, and S. Kraemer. 2005. The 10-minute oral presentation: What should I focus on? *Am. J. Med. Sci.* 329 (6):306–309.

Horiuchi, S., J.S. Nasser, and K.C. Chung. 2022. The art of a scientific presentation: Tips from Steve Jobs. *Plast. Reconstr. Surg.* 149 (3):533–540.

Kanpolat, Y., ed. 2002. *Research and publishing in neurosurgery*. Vol. 4. Vienna, NY: Springer-Verlag.

Katchburian, E. 2003. Oral presentations at scientific meetings: Some hints and tips. *Sao Paulo Med. J.* 121 (1):37–38.

Papanas, N., E. Maltezos, and M. Lazarides. 2011. Delivering a powerful oral presentation: All the world's a stage. *Int. Angiol.* 30 (2):185–191.

Roberts, L.W., ed. 2013. *Roberts academic medicine handbook: A guide to achievement and fulfillment for academic faculty*: Springer, New York.

Rougier, N.P., M. Droettboom, and P.E. Bourne. 2014. Ten simple rules for better figures. *PLoS Comput. Biol.* 10 (9):e1003833.

Storz, C. 2002. *Oral presentation skills–a practical guide*. Institut National de Télécommunications: Evry France.

Vučković-Dekić, L. 2002. Oral presentation. *Arch. Oncol.* 10 (3):212–213.

Wellstead, G., K. Whitehurst, B. Gundogan, and R. Agha. 2017. How to deliver an oral presentation. *Int. J. Surg. Oncol.* 2 (6):e25.

22 Research Grant Proposals

22.1 RESEARCH GRANT PROPOSALS *VERSUS* RESEARCH PAPERS

Both research grant proposals and research papers are research documents compiled by researchers using a well structured and connected thought process. Career advancement in the academia and industry (*i.e.*, research and development scientists) depends largely on *wining* research grants and *publishing* outcomes of the research. Therefore, it is imperative that academics as well as research and development scientists develop the skills necessary for writing successful research grant proposals (to obtain research funds) and for writing publishable papers (disseminating research outcomes). Although a few similarities exist between these documents, they are essentially different (Table 22.1).

Fundamentally, a research grant proposal is a *sales document* (selling an idea to a funder) while a research paper is a *disseminating document* (disseminating what has been learned from a research). In other words, researchers use grant proposals to acquire money to carry out research and then use research papers to disclose the outcomes of the research. Consequently, the language in grant proposals is *persuasive* while that in research papers is *explanatory* as noted by (Myers, 1991):

> "In research grant proposals, one must persuade without seeming to persuade; yet, every sentence is meant to persuade".

TABLE 22.1
Differences between a Research Paper and a Research Grant Proposal

Research Paper	Research Grant Proposal
Explanation and description of completed work	Descriptive plan of future work
Written for publication of findings of research	Written to obtain money and do research
Audience are mainly experts of the field	Audience is a mixture of experts and nonexperts
Language style is explanatory and impersonal (verbosity rewarded)	Language style is persuasive and personal (brevity rewarded)
Specialist terms are encouraged	Specialist terms are discouraged (simple and easily understandable language encouraged)
Contains title, author details, abstract, introduction, methodology, results, discussion, conclusions, acknowledgments and references	Contains title, applicant details, abstract, introduction, methodology and references. Does not contain results, discussion, conclusions and acknowledgments
Does not contain a budget and a timeline	Contains a budget and a timeline
Does not have a submission deadline	Has a submission deadline
Appraised mainly on the basis of knowledge contribution, agreement of conclusions with results	Appraised mainly on the basis of novelty of idea, significance of outcomes and credibility of applicant

DOI: 10.1201/9781003186748-22

Thus, the target of every research grant proposal is to persuade the reader that the:

- Proposed research has a clear research question.
- Proposed research has clear objectives.
- Proposed research is worth funding, *i.e.* – it will make a significant, original and important contribution to the knowledge base of the field.
- Objectives can be achieved with the proposed methods.
- Plan for achieving the objectives is well-thought through.
- Objectives can be achieved within the available timeframe.

The grant applicant persuades the reader on each of these points in different sections of the proposal as shown in Section 22.2.

22.2 COMPONENTS OF A RESEARCH GRANT PROPOSAL

Although funders normally state in their guidelines what to include in their grant proposals, including the font size and type to use, grant proposals commonly contain:

i. The research title or topic.
ii. Name(s) of applicant(s) and their addresses.
iii. Synopsis, abstract or summary of the research.
iv. Introduction.
v. Background and literature review.
vi. Methodology.
vii. Expected outcomes.
viii. Budget.
ix. Timeline.
x. References.

The names of these sections may vary across funders, but their meaning and what they are expected to contain are the same. For example:

- Problem/need/situation description.
- Work plan/specific activity.
- Outcomes/impact of activities.
- Impact evaluation.
- Budget justification/narrative.

22.3 WINNING A RESEARCH GRANT

Research grants are won because:

i. The applicant meets all the eligibility criteria.
ii. The research proposed is in line with the objectives/vision of the funder.
iii. The research proposed merits funding.
iv. The proposal is prepared in line with the guidelines (font type and font size, word limit, budget, submission deadline, *etc.*) of the funder.

22.3.1 ELIGIBILITY CRITERIA

Eligibility criteria (Figure 22.1) are conditions stipulated by a funder for applicants to meet before applying for their funding. These conditions vary from funder to funder, but they are prerequisites for grant applications. Grant proposals of applicants that do not meet all the eligibility criteria are normally not considered by funders whatsoever. Therefore, be sure that you meet all the eligibility conditions before submitting your grant application.

22.3.2 FUNDER OBJECTIVES

After fulfilling all the eligibility criteria, funders expect the proposed research to be in line with their objectives. Research proposals with objectives different from those of the funders are normally not funded whatsoever. For example, if the objective of a funder is to fund research projects that tackle the problem of deforestation in *Nigeria*, the funder will not fund a research proposal aimed at tackling deforestation in *Ghana*.

Applicant Eligibility

∞ Applicants must have a PhD or be in the final stage of their PhD provided that it will be completed (including viva) before the start date of the fellowship. Confirmation of award of the PhD will be required before any fellowship award is confirmed.

∞ Applicants should have no more than seven years active full-time postdoctoral experience at the time of application, including teaching experience, time spent in industry on research, honorary positions and/or visiting researcher positions. Career breaks must be clearly detailed and explained in the application, for example "Start and end dates - career break – maternity/paternity leave".

∞ Applicants should be working outside the UK and should not hold UK citizenship at the time of application.

∞ Applicants who are not currently employed are still eligible but will need to provide details of their previous supervisor.

∞ Individuals already living, working or undertaking research in the UK are **not** eligible to apply.

∞ Individuals who have lived, worked or undertaken research in the UK in the 12 months prior to the application deadline are **not** eligible to apply.

∞ Individuals working outside the UK but employed by a UK organisation are also not eligible to apply.

∞ Applications from individuals who have not studied or worked in the UK previously are encouraged as the scheme aims to establish new links between the applicant and the UK.

∞ Applicants who completed their PhD at a UK organisation must have been working and based outside the UK for **at least one year** at the deadline for the application.

∞ Applicants proposing to return to their UK-based PhD organisation and/or PhD supervisor or to their Post-doctoral supervisor will normally be considered to be ineligible and so applicants must have exceptional reasons for proposing to do so.

∞ Proposed Fellowships must be carried out in the UK at the UK host organisation for the duration of the Fellowship.

∞ Applicants must be competent in oral and written English. The applicant must confirm their competency on the application form and the UK Co-applicant needs to include the applicant's competency in their supporting statement.

∞ Individuals who have previously been in receipt of a Newton International Fellowship are not permitted to apply again.

∞ Applicants who have been unsuccessful in a previous round of the competition may make another application in this round.

FIGURE 22.1 Example of eligibility criteria of a funder (the Royal Society of Chemistry). Taken from the Newton International Fellowship Scheme notes (2020).

22.3.3   FUNDING MERIT OF RESEARCH

Funders fund only research that merit funding. "Merit" as used here means novel ideas that (i) can significantly impact the real world or the scientific world for better and (ii) are in line with their funding objectives/foci. As a research grant applicant, you can improve your chances of winning by submitting a novel idea that falls into the same category with those previously funded by the funder. For example, a funder that previously funded many computational chemistry research is unlikely to fund research in microfluidics and *vice versa*.

22.3.4   PROPOSAL PREPARED ACCORDING TO GUIDELINES

Funders give guidelines for preparing research grant proposals which applicants are expected to follow strictly. These guidelines range from text font type, font size, word limit and budget limit to submission deadline. Proposals exceeding the stipulated word and budget limits, prepared with a different font type and size, and also submitted late never get funded no matter the novelty of the research idea.

22.4   WRITING A RESEARCH GRANT PROPOSAL

The preparation of a research grant proposal is a five-stage process.

i. *Idea conception*: Research ideas can be conceived by critically examining the real world or the scientific world. Critical examination of the environment (air, water, soil), plants, animals and humans can lead to researchable ideas in terms of the real world. Examining the scientific literature of a subject area helps in identifying issues that require further investigation and that can be conceived into research ideas.
ii. *Research question formulation*: A research question or hypothesis is formulated from a research idea *vis-á-vis* the aim or the bigger picture (*i.e.*, the outcome of the research).
iii. *Formulation of objectives*: Objectives are formulated around the aim of the research. The outcome of each objective is a step toward achieving the aim of the research.
iv. *Methodology adoption*: The methods for achieving each objective are adopted.
v. *Proposal preparation*: The various sections (Section 22.2) of the proposal are written, with stages (i)–(iv) stated here as the framework. A summary of the contents of these sections and a plausible order for writing them are given in Figure 22.2. However, suggestions for writing them are given in Chapters 23–31.

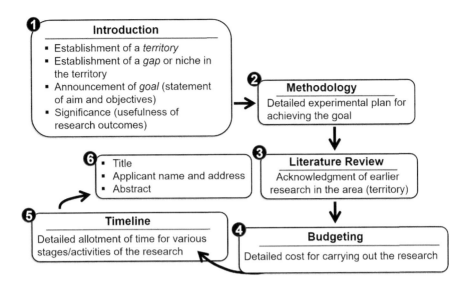

FIGURE 22.2 Schematic illustration of a plausible order for preparing a research grant proposal. Based on this order, the introduction, methodology, literature review, budget, timeline, title, applicant name and address and the abstract are written sequentially.

FURTHER READING

Open secrets about writing successful grant proposals: Notes for researchers and research managers rm_notebook_02.pdf (sarima.co.za)
- Contains general useful tips for writing successful grant proposals.

REFERENCE

Myers, G. 1991. *Writing biology: Texts in the social construction of scientific knowledge*: University of Wisconsin Press, Madison, WI.

23 Grant Proposal Title and Applicants' Details

23.1 FUNCTION OF A GRANT PROPOSAL TITLE AND HOW TO WRITE A GRANT PROPOSAL TITLE

The title of a grant proposal is a one-sentence summary of what the proposal is all about; therefore, it should precisely capture the proposal's essence. Just like the titles of scientific papers (Chapter 3), grant proposal titles may be descriptive, declarative or interrogative; thus, the guide given in Section 3.4 on writing the titles of scientific papers applies here too. However, while there is no limit to the degree of field-specific jargons in the titles of scientific papers, grant proposal titles are expected to contain a few field-specific jargons as these proposals are reviewed by a mixture of both experts and nonexperts of the proposed research (Goldbort, 2008). Because the title occupies a prominent part of the grant proposal and because reviewers read it before any other part of the grant proposal (Smith et al., 2009), grant proposal titles have both a visual and a cognitive influence on the reader. A concise and an enticing title may invite a reviewer to read a particular proposal first before reading others. Additionally, such a title gives the reviewer a first good impression of the overall proposal (Marshall, 2013) and also reveals the applicants' ability to express ideas and thoughts clearly to both experts and nonexperts. Therefore, it is imperative to formulate a concise, simple and an enticing title that precisely indicates the essence of your grant proposal to both an expert and a nonexpert. I found some research grant proposal samples online with excellent titles (see Appendix III). The titles of the first two proposals are "Crystal Growth in Open Framework Inorganic Materials" (Grant Proposal I) and "Assessing the Roles of Biofilm Structure and Mechanics in Pathogenic, Persistent Infections" (Grant Proposal II). The first is an Engineering Physical Science Research Council grant awarded to Prof. Michael William (principal investigator) and others (1998). The second is a National Institute of Allergy and Infectious Diseases grant awarded to Dr Vernita Gordon and others (2017). These titles have clearly and concisely summarized the essence of the grant proposals without any ambiguity.

23.2 DETAILS OF APPLICANT(S)

Just like research papers, an individual can submit a grant proposal alone or with others (co-applicants). Thus, this section provides the names, qualifications and addresses (institutional or industrial) of all the applicants. If submitting with co-applicants, the lead applicant, investigator or researcher (*i.e.*, the person the success of the proposed research strongly depends on) must be clearly indicated. Some funders stipulate specific qualities (qualifications, age, nationality, *etc.*) for the lead applicant. To avoid an unsuccessful application, be sure that the supposedly lead applicant meets all the qualities before applying.

DOI: 10.1201/9781003186748-23

FURTHER READING

Denscombe, M. 2012. *Research proposals: A practical guide:* McGraw-Hill Education (UK).
  • Contains useful tips for writing various sections of a research grant proposal including the title.
Ohtake, P.J. 2000. Research corner: Grant writing: Toward preparing a successful application. *Cardiopulm. Phys. Ther. J.* 11 (2):69–73.
  • Discusses essential aspects of research grant proposals including the title.

REFERENCES

Goldbort, R. 2008. *Writing for science*: Yale University Press. Binghamton, New York.
Marshall, L.S. 2013. Research commentary: Grant writing: Part II grant application/proposal components. *J. Radiol. Nurs.* 32 (1):48–51.
Smith, G.F., E. Figueiredo, T. Pennington, and P. Davila. 2009. Getting that grant: How to convince an evaluation panel that your proposal is worthy of funding. *TAXON* 58 (2):675–677.

24 Grant Proposal Abstract and Summary

24.1 ABSTRACT *VERSUS* SUMMARY

For a research paper, the abstract serves as a summary of the paper and reduces all the sections: introduction (*i.e.*, background, problem and aim), methods, results and conclusions into ≤300 words. The reduction of these sections is disproportional to their length in the paper, and thus the abstract is said to have a nonlinear relationship with them. Contrarily, for research grant proposals, the abstract does not serve as a summary of the proposal in that it reduces certain sections of the grant proposal and leaves out certain sections. The reduction of the sections included in the abstract is also disproportional to the actual sections in the proposal, *i.e.*, a nonlinear relationship exists between the abstract and the included sections. This can be compared with the summary (also called executive or project summary) which reduces all the sections of the grant proposal proportionately, giving rise to a linear relationship between the summary and the grant proposal itself (Goldbort, 2008). Funders generally ask for either the abstract or the summary, but some funders consider the abstract as the project summary. Be sure of what is being asked and do not to interchange between the abstract and the summary. If the funder asks for an abstract, be sure to supply it and vice versa if a summary is asked. For instance, if the funder asks for a 200-word project summary, know that the funder is asking for the abstract. However, if the funder asks for a one or a two-page project summary, know that the funder is asking for a summary of the entire proposal. Also, some funders specify the contents of their abstract and summary; be sure to follow these specifications meticulously. In some cases, funders ask for both the abstract and the project summary. In this case, the abstract is meant for the expert readers while the summary is meant for the nonexpert readers (Kaplan, 2012).

24.2 FUNCTION OF A GRANT PROPOSAL ABSTRACT AND SUMMARY

The abstract of a grant proposal serves as the proposal synopsis or table of contents. It is descriptive in nature and generally contains the following subsections:

a. The *background* of the proposed project, which states established knowledge that relates to the area of the proposed project.
b. A statement of the *problem*, which states the knowledge *gap* that needs to be filled or addressed.
c. The *aim* and the *objectives* of the proposed project, which state the specific thing you want to do to fill the knowledge gap (*aim*) and the things that will lead you to what you want to do or solution of the problem (*objectives*).

DOI: 10.1201/9781003186748-24

   d. Preliminary *data* (if available), which gives data from pilot experiments.
   e. The *methodology*, which describes the experimental methods that will be used in achieving the set objectives and, by extension, the aim.
   f. The *significance* of the proposed project, which states the importance or benefits of the project outcomes.
   g. The total *cost* of the project.

Additionally, evidence for the *credibility* of the applicants and their *ability* to successfully execute the project, a description of applicants' organization and a description of how the aim of the proposed project matches the goals of the funder can be optionally included in the abstract. However, unlike the structured abstracts of scientific papers where these subsections are clearly marked, these subsections are not clearly marked in grant proposal abstracts. These subsections overlap strongly with those in the introductory section of scientific research papers (Chapter 9) and grant proposals (Chapter 24). Ismaeel (2016) analyzed the abstract of 20 successful grant proposals of the National Institutes of Health (USA) and reported that they contain predominantly the *aim* or *objectives* of the proposed project, the *methodology* and the *cost* implication. This stresses the importance of these components in grant proposal abstracts. Ismaeel (2016) also noted that these abstracts either began with the aim (or objectives) or the cost implication. This is in line with what a program manager of a funder recommended: the first sentence of a grant proposal abstract should say "the research objective of this proposal is ..." (Kaplan, 2012). Comparatively, the summary of a grant proposal is a mini-version of the proposal, and thus contains an abridged version of all the sections of the proposal (Goldbort, 2008).

24.3   WRITING A GRANT PROPOSAL ABSTRACT AND SUMMARY

The abstract and the summary of a grant proposal are easier to write after all the other sections of the proposal have been written because a clearer picture of the proposal emerges at this point. The abstract or the summary is the second section, after the title, that reviewers are going to peruse to gain an overall impression of the proposed project and the applicants' ability to execute it successfully and satisfactorily (Lindgreen et al., 2019). The reviewers may not read the proposal further once they gain a negative impression about the project or the credibility of the applicants (Lindgreen et al., 2019). Be sure that your abstract or summary is error-free and can be understood without reference to any sections of the proposal. Typographical and grammatical errors as well as unexplained symbols and abbreviations will give the reviewers a bad impression about the proposal and the credibility of the applicants (Lindgreen et al., 2019). Because reviewers use the abstract or the summary to make preliminary decisions about the success of the project and the credibility of the applicants, keep the abstract or summary informative (in line with Section 24.2), persuasive, succinct and error-free (Kaplan, 2012). In fact, your entire proposal should be free of errors – formatting, grammatical and spelling errors (Wegener and Katan, 2018). The abstract or summary should also be understandable by nonexperts readers on the appraisal team. An example of a grant proposal abstract is given in Example 24.1, which can be compared with the project summary of a grant proposal in Example 24.2.

Example 24.1 *Content Analysis of the Abstract of a Successful Chemistry Grant Application*

The funder (Engineering Physical Science Research Council, EPSRC) said, "describe the proposed research using (about 200) words geared to the non-specialist reader".

DESCRIPTION (GRANT PROPOSAL I, APPENDIX III)

"We propose the most fundamental, ambitious and concerted multi-disciplinary investigation into the understanding of crystal growth and rational design of open framework materials yet attempted. We believe the findings from this study will mark a major leap forward into our understanding of crystal growth and our ability to exploit our understanding to produce new materials with unique properties and applications. Extensive studies on the synthesis of porous materials have been carried out. However, the majority of this synthetic work has been aimed primarily at either (i) the discovery of new structures, (ii) modification or improvement of existing materials or (iii) process development to enable such materials to be produced successfully on a large scale. The effort so far on synthesis and crystallization mechanism has yielded many positive results but also many unanswered questions, for example: (i) the detailed mechanism of nucleation (ii) the identity of growth species and (iii) whether nanocrystal growth occurs by addition or aggregation. This research involves the application of a powerful set of complimentary techniques to the study of crystal growth of open-framework materials comprising: atomic force microscopy, high resolution transmission and scanning electron microscopies, in-situ NMR with enhanced data processing, X-ray diffraction and mass spectrometry. An improved understanding of the synthesis process is likely to yield important economic benefits, for example, better process control, increased efficiency in reagent usage, improved reproducibility and the capacity to modify or tailor products for specific applications. Perhaps most important of all would be the ability to identify successful synthetic routes to as-yet unknown structures and compositions which have been predicted on theoretical grounds to have beneficial characteristics. Such a step forward to a new level of primary understanding would open the way to innovative applications in chemistry, physics (ordered arrays) and biomaterials".

From an EPSRC chemistry grant awarded to Prof. Anderson M. William (University of Manchester Institute of Science and Technology) and co-applicants (1998)

ANALYSIS OF ABSTRACT

Aim: We propose the most fundamental, ambitious and concerted multi-disciplinary investigation into the understanding of crystal growth and rational design of open-framework materials yet attempted. *Significance:* We believe the findings from this study will mark a major leap forward into our understanding of crystal growth and our ability to exploit our understanding to produce new materials with unique properties and applications. *Background:* Extensive studies on the synthesis of porous materials have been carried out. However, the majority of this synthetic work has been aimed primarily at either (i) the discovery of new structures, (ii) modification or improvement of existing materials or (iii) process development to enable such materials to be produced successfully on a large scale. *Gap* or *problem:* The effort so far on synthesis and crystallization mechanism has yielded many positive results but also many unanswered questions, for example: (i) the detailed mechanism of nucleation, (ii) the identity of growth species and (iii) whether nanocrystal growth occurs by addition or aggregation. *Method:* This research involves the application of a powerful set of complimentary techniques to the study of crystal growth of open-framework materials comprising: atomic force microscopy, high resolution transmission and scanning electron microscopies, in-situ NMR with enhanced data processing, X-ray diffraction and mass spectrometry. *Significance:* An improved understanding of the synthesis process is likely to yield important economic benefits, for example, better process control, increased efficiency in reagent usage, improved reproducibility and the capacity to modify or tailor products for specific applications. Perhaps most important of all would be the ability to identify successful synthetic routes to as-yet unknown structures and compositions which have been predicted on theoretical grounds to have beneficial characteristics. Such a step forward to a new level of primary understanding would open the way to innovative applications in chemistry, physics (ordered arrays) and biomaterials.

In line with the recommendation of Kaplan 2012, the grant proposal abstract given in Example 24.1 began with the *aim* followed sequentially by the *significance, background, gap* or *problem, method* and then ended with the *significance* (economic benefits) of the proposed research. Additionally, the contents of this abstract are in line with Sec. 24.2, excluding the cost implication, applicant credibility and capability as well as the applicant's organizational information and a matching of the project goals with those of the funder. Also, notice that the abstract is written in *future tense* as the work is yet to be done. Although the *simple present tense* is used with the collective pronouns "we" and "our" in few sentences. This is in contrast to the use of *simple present tense* or *simple past tense* in the abstracts of scientific research papers that report completed research work. Similarly, the project summary in Example 24.2 contains the *background, problem, aim* and *objectives* in the first paragraph. The second paragraph contains highlights of the methodology and potential significance of the results. Finally, the broader impacts of the project are highlighted in the last paragraph. The summary is also written in *future tense* using the passive voice. Overall, the abstract is much shorter than the summary. Additionally, unlike the abstract that serves as the proposal synopsis, the summary gives the reader a complete overview of the proposed project.

Activity 24.1 *Analyzing a Grant Proposal Abstract*

Using Section 24.2, analyze the following bioengineering grant proposal abstract into the various subsections.

DESCRIPTION (GRANT PROPOSAL II, APPENDIX III)

"What spatial structure and mechanics develops in biofilm infections, and how such spatial structure and mechanics impact the persistence and virulence of biofilm infections, is not known. The long-term goal is to find diagnostic and treatment approaches that address the structure and mechanics of multicellular, three-dimensional biofilm infections within the host. The objective of this application is to determine the mechanics and structure of biofilm infections of the opportunistic pathogen *Pseudomonas aeruginosa* in chronic wounds, and how these physical properties impact disease course. The central hypothesis is that spatial structure and mechanics are the major physical factors controlling virulence, antibiotic resistance, and immune evasion in biofilm infections. The rationale underlying this application is that completion will identify key physical targets for preventing, disrupting, or ameliorating biofilm infections for an important biofilm-forming pathogen. The proposed work will also develop a widely-applicable platform for assessing the state and impact of biofilm structure and mechanics for other infecting organisms. The central hypothesis will be tested by pursuing three specific aims: 1) Determine the spatial structure and mechanics of in vivo biofilm infections; 2) Determine how spatial arrangements differentiate into distinct microenvironments; 3) Determine the role of spatial structure and mechanics in biofilm-neutrophil interactions. We will pursue these aims using an innovative combination of analytical and manipulative techniques from both biological and physical sciences. These include both recently-developed techniques specific to biofilm studies, and more-established techniques that have been applied very little to the study of biofilm materials. The proposed research is significant, because it will determine which structural and mechanical characteristics should be therapeutic targets. It is also significant because it will develop a platform that can be extended to study other pathogens (or commensals) and synergies to open new avenues for biofilm therapies. This work will develop foundational resources that will be used by other researchers, for *P. aeruginosa* and other organisms. The proximate expected outcome of this work is an understanding of which biofilm structural and mechanical characteristics contribute to clinical impact. The results will have an important positive impact immediately because they will establish better understanding of biofilm infection, virulence, and resistance to antibiotics and the immune system for an important pathogen, and long-term because they lay the groundwork to develop a suite of techniques for better treatment of biofilm infections".

From a National Institutes of Health bioengineering grant awarded to Dr Vernita Gordon (University of Texas, Austin) and co-applicants (2017)

Activity 24.2 *Identifying Signal Words in a Grant Proposal Abstract*

Identify the words that signal the aim or objective, background, gap or problem and significance in the grant proposal abstracts shown in Example 24.1 and Activity 24.1.

Example 24.2 *The Summary Section of a Chemistry Grant Application*

PROJECT SUMMARY (GRANT PROPOSAL III, APPENDIX III)

Background: Thermal expansion is an important materials property from the standpoint of engineering applications, as mismatches in thermal expansion can result in stress, cracks, or separation at interfaces. **Problem:** It is therefore desirable to control and, in many cases, minimize the thermal expansion of materials. **Aim:** This CAREER proposal is aimed at establishing a basis for the integration of research and education that will lead to the development of materials and composites that can overcome expansion related problems in many applications. Negative thermal expansion (NTE) materials are particularly promising for use in composites, as they are expected to show a more pronounced effect on the composite at equal loading. **Objectives:** This proposal seeks to achieve the following specific objectives: (1) To gain a fundamental understanding of factors that influence the expansion and phase transition behavior of NTE materials through the preparation and characterization of new NTE compounds, (2) to characterize the high-pressure behavior of these materials, (3) to prepare NTE/polymer composites with tailored thermal expansion properties, (4) to integrate low temperature approaches used for the synthesis of NTE and other metastable materials into the undergraduate and graduate level chemistry curricula, (5) to realize the participation of undergraduates and high school students in this research, and (6) to enhance the public image of materials chemistry through outreach activities. **Significance:** The proposed research will *intellectually* contribute to the scientific literature through the preparation and characterization of new NTE materials belonging to the $Sc_2W_3O_{12}$ family. **Methodology:** Non-hydrolytic sol-gel routes will be used in the synthesis. This will allow incorporation of cations into the framework that are not accessible by ceramic methods, and straightforward preparation of mixed cation compounds. Analysis by variable temperature powder X-ray diffraction combined with Rietveld refinement will reveal factors that influence the occurrence of temperature-induced phase transitions and the materials' expansion behavior. **Significance:** These results will enable researchers to predict the behavior of both pure and mixed cation systems. **Methodology:** The high-pressure stability of the compounds will be established by Raman spectroscopy and synchrotron diffraction studies in a diamond anvil cell. **Significance:** A *second intellectual* contribution will be related to the exploration of NTE/polymer composites in film and fiber form. **Methodology:** Special attention will be directed towards the interface region between polymers and oxide particles. The surface of the oxide will be modified by grafting of organic groups that will result in favorable interactions or copolymerization with the polymer precursors. Interface interactions, expansion properties, and changes in mechanical properties of the composites will be characterized. **Broader Significance/Impact:** The *first broader impact* of this proposal will be the integration of advanced materials and low temperature methods used for their preparation into the graduate and undergraduate curricula of the chemistry program at the University of Toledo through the design of a new course in solid-state chemistry. The *second broader impact* lies in the inclusion of powder diffraction methods and analytical tools for powder data ranging from indexing to Rietveld refinement into the existing graduate and undergraduate level Crystallography course. The principal investigator also shares in the responsibility for training researchers on the Department of Chemistry's state-of-the-art powder diffractometer, making her knowledge available to the scientific community at the University of Toledo and local industry. She regularly assists researchers in a variety of standard and non-standard powder diffraction experiments. A *third broader impact* will result from the exposure of undergraduate and high school students to scientific research on projects appropriate for their skill levels. This experience is likely to stimulate their interest in science and will prepare them for choosing a career path. The *final broader impact* will be the enhancement of the public recognition of materials chemistry through outreach activities, using the intriguing concept of materials that shrink when heated, and how their use in composites can help overcome "real-life problems" related to thermal expansion.

From a National Science Foundation grant awarded to Prof. Cora Lind-Kovacs (University of Toledo, Ohio)

FURTHER READING

Glasser, S.P., and P. Glasser. 2008. *Essentials of clinical research.* Springer.
- Emphasizes the centrality of grant proposal abstracts and offers suggestions for writing a grant proposal abstract.

Goldbort, R. 2008. *Writing for science.* Yale University Press. Binghamton, New York.
- Discusses the summary section of a research grant proposal and contains an example of a grant proposal summary drawn from a proposal submitted to the National Science Foundation.

REFERENCES

Goldbort, R. 2008. *Writing for science*: Yale University Press. Binghamton, New York.
Ismaeel, Y. 2016. "Why Do Some Grant Proposals Win the Funds and Others Fail? Genre Analysis for the Abstracts of Grant Proposals Submitted by the Health Science Center Researchers to the National Institutes of Health." MSc Thesis, West Virginia University.
Kaplan, K. 2012. Funding: Got to get a grant. *Nature* 482 (7385):429–431.
Lindgreen, A., C. Anthony Di Benedetto, C. Verdich, J. Vanhamme, V. Venkatraman, S. Pattinson, A.H. Clarke, and Z. Khan. 2019. How to write really good research funding applications. *Ind. Mark. Manag.* 77:232–239.
Wegener, S., and M. Katan. 2018. Getting the first grant. *Stroke* 49 (1):e7–e9.

25 Research Proposal Introduction

25.1 FUNCTIONS OF THE INTRODUCTORY SECTION OF A RESEARCH GRANT PROPOSAL

The introductory section of a research grant proposal serves as the framework for the proposed research, and it contains (Al-Riyami, 2008):

- A brief *background* (*i.e.*, the setting or foundation) of the proposed research. This establishes the *territory* (or context) of the proposed research.
- A *statement of problem*, establishing the *gap* (or niche) in the territory of the research.
- Specific *aim* (or goal) of the proposed research, stating how the established gap will be filled.
- *Objectives* of the proposed research, stating various tasks necessary for achieving the specific aim.
- *Significance* of the proposed research, stating the *usefulness* (or benefits) of anticipated results or outcomes.
- Optionally, a *competency claim* (*i.e.*, a summary of research track-record of the applicant on the subject area).

25.2 WRITING THE INTRODUCTORY SECTION OF A RESEARCH GRANT PROPOSAL

25.2.1 ESTABLISHING THE TERRITORY/CONTEXT

A research grant proposal can be situated in either three territories (Connor and Mauranen, 1999):

i. *A research field territory*: *i.e.*, the discipline(s) the research can be identified with.
ii. *A "real-world" territory*: *i.e.*, the real-life situations the research can be identified with (*e.g.*, environment, food, medicine, *etc.*)
iii. *Or a combination of research field territory and real-world territory*: *i.e.*, the research field and the real-life situations the research can be identified with. Excerpts of research grant proposals, from the University of California Irvine, containing examples of these territories are shown in Example 25.1.

DOI: 10.1201/9781003186748-25

Example 25.1 *Territories of Research Grant Proposals*

Excerpts from the background subsection of the introductory section of research
grant proposals (Appendix III), with research field, real world and a combination
of research field and real-world territories.

Background	Territory
Grant Proposal IV	
"Nanoscience is a growing field of research merging chemistry, physics and biology, with sweeping implications for new technologies. One topic of interest within this field is the electronic properties and applications of nanowire materials. Researchers have primarily focused on making interesting circuits with these 'wires' and the work was named 'Breakthrough of the Year 2001' in science magazine".	Research field
Grant Proposal V	
"Strain (deformation of a material) is an important measure for studying osteoporosis, tumors in bone and designs of joint prostheses. However, measurement of strain on the surface of bone with high fidelity has been extremely difficult due to the lack of suitable tools. Current strain gauges are relatively large (2 mm by 5 mm gauge) and are difficult to mount on bone. As a result, strain gauges are not used very often, and when they are used, the fewest possible are used. In addition, when one uses relatively few gauges, one must know where to put the gauge if one wants to measure strain at the site of maximum or minimum strain. Because the strain gradients in bone can be extreme, mounting a gauge just a few mm or a cm away from the peak strain (or minimum strain) can lead to grossly understated (or overstated) measured values. Finally, with a gauge measuring 5 mm in length, the local strain cannot be measured and only an average strain in the region covered by the gauge is measured".	Real world
Grant Proposal V	
"Increasing energy demands and depletion of fossil fuel resources worldwide demands the immediate attention of science in general and chemistry specifically to develop alternative sources of reliable energy. At our current rate of consumption, 500-million-year-old fossil fuel sources will run dry in less than 100 years. Carbon dioxide emissions are at an all-time high and are projected to increase at an alarming rate by 2100. Our future security and even the sustainability of life depend on a change in these global energy trends".	Research field and real world

Activity 25.1 *Naming Grant Proposal Territories*

Name the specific territories (*e.g.*, medicine, biology, physics, *etc.*) of the grant
proposals described in Example 25.1.

Notice from Example 25.1 that a research proposal is situated in a territory by progressively moving from general statements to specific statements, *i.e.*, broadest → broad → specific statements.

25.2.2   ESTABLISHING GAP IN THE TERRITORY

The purpose of a problem statement is to create dissonance between the present situation and the ideal situation, *i.e.*, what ought to have been. In other words, it says "things are like this", but "they ought to have been like this". With these, the gap in the territory is established. Discussing the present requires a review of recent relevant literatures, events or trends. Discussing the ideal situation may require citing literatures that support the situation, mentioning the goals of the funder or the benefits of the situation. The gap in the territory serves as the motivation for the proposed research (Connor and Mauranen, 1999). Similarly, the gap can either be in:

  i. A research field territory.
 ii. A real-world territory.
iii. Or a combination of research field and real-world territories (Connor and Mauranen, 1999) as illustrated in Example 25.2.

Example 25.2 *Establishing Gaps in Research Grant Proposals*

Examples of gaps in a research field, a real world and a combination of research field and real-world territories in research grant proposals (Appendix III).

Gap	Territory
Grant Proposal IV	
"The research group of Professor Collins currently grows high quality carbon nanotubes by chemical vapor deposition and incorporates them into nanoelectronic circuits for testing. The focus and expertise of the group is the interaction of these circuits with different environments and the potential sensor applications. (Gap) However, numerous problems exist with such circuits. The nanotubes are weakly sensitive to chemicals all along their length, with most of the sensitivity concentrated at point defects in the nanotube structure. At present, these defects happen by chance in random positions, making it difficult to further optimize them for applications".	Research field
Grant Proposal V	
"When measuring strain to evaluate a prosthesis design, localized changes in strain are thought to be key indicators that potentially destructive bone remodeling will occur (Keyak, 2001). For example, if strain decreases upon implantation of a prosthesis, bone will remodel to become less dense (bone resorption) and prosthesis failure can result. Bone resorption is a major clinical problem in orthopedics. On the other hand, if strain increases, the patient can experience pain and bone hypertrophy can result. This situation occurs near the distal tip of the femoral component of a hip prosthesis (Skinner et al., 1994; Lee et al., 1994; Xu et al. 2003). (Gap) Currently available strain gauges are too large and too difficult to handle to adequately measure the changes in strain upon implantation of a prosthetic".	Real world

(Continued)

Gap	Territory
Grant Proposal VI	
"In nature, green plants harness sunlight to drive the conversion of CO_2 and H_2O into O_2 and simple carbohydrates. An attractive artificial photosynthetic strategy could harness sunlight to split water into H_2 and O_2 *via* multi-electron reactions. (Gap) While sunlight may be used to provide the thermodynamic driving force for these processes, synthetic molecules capable of promoting the specific bond-breaking and bond-making reactions do not exist".	Research field and real world

25.2.3   Specific Aim – Filling the Gap

The specific aim of the research proposal addresses how the gap identified will be filled. This is done by either (ii) announcing what the research is all about or (ii) stating the research question or hypothesis. The specific aim can be research field-related, real-world-related or a combination of both, depending on the formulation as shown in Example 25.3.

Example 25.3 *Specific Aim of Research Grant* Proposals

Examples of the specific aim of a research grant proposal (Appendix III), announcing the filling of an identified gap.

Specific Aim	Territory
Grant Proposal IV	
"In traditional microelectronics, one way to concentrate a particular effect at a known location is to form a junction between two different materials. For example, a diode is a junction between p-doped semiconductor and n-doped semiconductor. The special properties of the junction allow a diode to detect light (a photodiode) or emit light (an LED), even though the different semiconductors themselves only have weak optical properties. (**Aim**) By incorporating similar junctions into our nanotube devices, we aim to build similar improvements into our chemical detectors".	Research field
Grant Proposal V	
(**Aim**) "Ultimately the goal is to develop a wireless, implantable array that can be attached to the bone's surface. The array will be embedded in a flexible polymer membrane (a "skin"), enabling real-time data logging of strain on the bone surface".	Real world
Grant Proposal VI	
(**Aim**) "This research proposal will develop new transition metal complexes and explore fundamental reaction chemistry related to the problem of photochemical hydrogen and oxygen production from aqueous solution".	Research field and real world

25.2.4 OBJECTIVES

The objective subsection contains a list of various tasks necessary for achieving the specific aim. Generally, 3–5 objectives are sufficient. The objectives should be clearly defined, achievable and should be targeted at achieving the specific aim.

25.2.5 SIGNIFICANCE – USEFULNESS OF RESEARCH

Results or outcomes of the proposed research can be useful to the research field, real world or a combination of both (Example 25.4).

Example 25.4 *Significance of Proposed Research*

Examples of usefulness of proposed research to the research field, real world or both research field and real world (Appendix III).

Significance	Territory
Grant Proposal VII	
"The production of a mineral precipitate and a decrease in the calcium levels in solution would indicate that phage-mediated lysis of cyanobacteria could be contributing to the formation of microbialites in the Great Salt Lake. The purpose of this experiment is to discover if lysis of these photosynthetic bacteria can cause spontaneous precipitation of calcium carbonate. This would be an important discovery. If we do see calcium carbonate precipitating, it may bring us one step closer to understanding the mystery of how these microbialites form".	Research field
Grant Proposal V	
"Through this research we will ultimately develop an implantable device to provide high-resolution mechanical data from live bone in real-time. With this device, physicians can diagnose bone disorders at the early stages where it can be corrected using relatively non-invasive procedures that patients can easily recover from. Moreover, after major bone surgery such as bone tumor removal or the insertion of a prosthetic, scientists and physicians will be able to monitor how the bone is healing after the surgery. Bone surface strain is a strong indicator of the bone's response to mechanical loading, and thus provides an important post-treatment metric of the bone's recovery. The device that we are developing will have a major impact on the diagnosis and treatment of bone disease".	Real world
Grant Proposal VIII	
"This project will benefit our understanding of enzymes and their role in disease, as well as improve our lab techniques and project planning. This experience will be valuable for both of us as we move forward in our education and future careers in science".	Research field and real world

From the foregoing, you may have noticed that the introductory section of a research grant proposal is very similar to that of a scientific research paper (Chapter 9) as they both contain the same elements and language of presentation.

FURTHER READING

Charles, M., S. Hunston, and D. Pecorari. 2009. *Academic writing: At the interface of corpus and discourse*. Bloomsbury Publishing.
- Contains an analysis of rhetorical patterns in research grant proposals, *i.e.*, it uncovers the style used by researchers in writing the various sections of research grant proposals.

Denscombe, M. 2012. *Research proposals: A practical guide:* McGraw-Hill Education (UK).
- Contains useful suggestions for writing the introductory section of a research grant proposal.

Koppelman, G.H., and J.W. Holloway. 2012. Successful grant writing. *Paediatr. Respir. Rev.* 13 (1):63–66.
- Contains a comprehensive discussion of the essential elements of the introductory section of research grant proposals.

Monte, A.A., and A.M. Libby. 2018. Introduction to the specific aims page of a grant proposal. *Acad. Emerg. Med.* 25 (9):1042–1047.
- Contains useful suggestions for writing the specific aims/objectives of a research grant proposal.

REFERENCES

Al-Riyami, A. 2008. How to prepare a research proposal. *Oman Med. J.* 23 (2):66–69.
Connor, U., and A. Mauranen. 1999. Linguistic analysis of grant proposals: European Union Research Grants. *English Specif. Purp.* 18 (1):47–62.
Keyak, J.H. 2001. Improved prediction of proximal femoral fracture load using nonlinear finite element models. *Med. Eng. Phys.* 23 (3):165–173.
Lee, I.Y., H.B. Skinner, and J.H. Keyak. 1994. Effects of variation of prosthesis size on cement stress at the tip of a femoral implant. *J. Biomed. Mater. Res.* 28 (9):1055–1060.
Skinner, H.B., D.J. Kilgus, J. Keyak, E.E. Shimaoka, A.S. Kim, and J.S. Tipton. 1994. Correlation of computed finite element stresses to bone density after remodeling around cementless femoral implants. *Clin. Orthop. Relat. Res.* 305:178–189.
Xu, Y., Y.-C. Tai, A. Huang, and C.-M. Ho. 2003. IC-integrated flexible shear-stress sensor skin. *J. Microelectromech. Syst.* 12 (5):740–747.

26 Background and Literature Review of a Research Grant Proposal

26.1 FUNCTIONS OF THE BACKGROUND AND LITERATURE REVIEW SECTION OF A RESEARCH GRANT PROPOSAL

The "Background and Literature Review" section is also called "Background and Justification", "Literature Review" or "Background". This section must not be confused with the background in the introductory section in Chapter 25 that establishes the territory of the research. The functions of this section are (i) to set the scene for the proposed research (*background*) and (ii) to show how the proposed research is related to earlier research works in the area (*literature review*) (Wiseman et al., 2013, Zlowodzki et al., 2007).

26.2 WRITING THE BACKGROUND AND LITERATURE REVIEW SECTION OF A RESEARCH GRANT PROPOSAL

a. *Background*

Reiterate the research problem and state the accomplishments of previous researches in relation to the problem. This will naturally lead to the review of relevant literatures.

b. *Literature review and justification*

The literature review should (i) focus on only relevant literatures and (ii) should include your work (to show competence) and those of other experts (to acknowledge their contributions) (Wiseman et al., 2013). The literature review should create a strong dissonance that justifies the need for your research. This will increase your chances of getting funded.

Example 26.1 *The Background Section of a Successful Grant Proposal*

BACKGROUND (GRANT PROPOSAL I, APPENDIX III)

"This proposal concerns crystal-growth mechanisms in open-framework materials. Our approach is multi-disciplinary involving the world's leading experts in the study of a generic problem. To appreciate its importance and timeliness and why such substantial investment should be made by EPSRC, it is necessary to understand the history of the subject, the role of UK science therein and the potential wide-ranging benefits.

DOI: 10.1201/9781003186748-26

HISTORY

Solid-state materials fall into a number of different categories differentiated either in terms of chemistry (*i.e.*, structure, bonding, crystallography, composition, *etc.*) or in terms of properties (*i.e.*, applications or potential applications). In chemical terms, one very large class of materials with many interesting and important properties is that of open-framework structures. What sets this class of materials apart from other materials is the potential (often unrealized) for highly ordered porosity whereby the entire solid material can be tangibly accessed by guest molecules. Such a highly desirable property – for many applications from catalysis to separations to sensors – is unique to this class of material. The archetypal material usually chosen to typify this class is the zeolite (crystalline aluminosilicate); however, a wide variety of framework compositions now exist including both fully inorganic and inorganic/organic hybrid networks.

How crystals nucleate and grow is a problem that has challenged scientists for many years (Wulff, 1901; Gibbs, 1928; Burton, 1951). How order is created from disorder, the driving forces involved, the quest for crystalline perfection. In many ways, nucleation and crystal growth should not be considered as separate phenomena; however, for practical reasons it is useful to do so. The techniques at our disposal to follow the nucleation and crystal-growth stages are substantially different, and therefore, there is often a visible seam between our perception of the two processes. In terms of crystal growth, the advent of scanning probe microscopies (SPM) (Binnig et al. 1982) and in particular atomic force microscopy (AFM) (Binnig et al., 1986) has permitted the detailed observation of nanometer-sized events at crystal surfaces. This is often possible under in-situ crystal-growth conditions as the technique can be operated to observe surfaces under solution. Real-time images of growing crystals have revealed terrace growth, spiral growth, the inclusion of defects and the occlusion of foreign particles in a wide variety of growth studies (McPherson et al., 2000). The effect of altering the growth medium on individual growth processes can be studied, for instance by adding proteins during the growth of abalone. Enhanced or retarded growth rates which result in altered morphology can be observed (Zaremba et al., 1996). By measuring real-time micrographs at a range of temperatures, the free energy for individual growth processes can be determined. To date, most of these crystal-growth studies have been on dense phase ionic crystals, such as calcite, or molecular crystals, such as proteins and viruses. There has been a modest amount of work performed on open-framework crystals such as zeolites and zeotypes of which we have been at the forefront (Anderson et al., 1996; Agger et al., 1998; Anderson et al., 2001; Singh et al., 2002; Agger et al., 2003; Dumrul et al., 2002; Bazzana et al., 2002). The reason for this is two-fold: first, often the most interesting open-framework structures can only be crystallized as micron-sized crystals, making observation by AFM a little more challenging; second, there has been a recent emphasis within the community on making new materials rather than on understanding formation. In our view, this is an oversight which is clear by the vast amount of new information forthcoming on understanding crystal growth in macromolecular systems which is helping address problems such as overcoming crystal size limitations, improving crystal purity, controlling intergrowth structures and controlling crystal habit. In open-framework materials, a better understanding of the crystal-growth processes will lead to new methodologies to control similarly important crystal features. But furthermore, it could lead to both new structures and also more cost-effective routes to existing but prohibitively expensive known structures.

THE ROLE OF UK SCIENTISTS

The UK has been at the forefront of utilizing atomic force microscopy in the study of fundamental crystal-growth processes in framework materials through work at UMIST (Anderson, Agger). Also, the teams at the Royal Institution (Slater), UCL (Lewis) and at Bath University (Parker) are leading in the use of modeling techniques to study energetics at the surfaces of framework materials. Furthermore, there is a wide literature concerning crystal growth of framework materials *via* conventional means utilizing optical microscopy and particle-size counting in highly controlled environments. Cundy, at UMIST, worked in this area at ICI for some 20 years and has an extensive knowledge of the lessons learned from this wealth of data enabling new results to be considered in the light of previous knowledge. Although AFM and theoretical studies are two of the most important recent routes to probe crystal-growth pathways, they must be supplemented by other techniques, in particular: high-resolution electron microscopy; NMR; diffraction; more recently mass spectrometry. The UK has not been at the forefront in this regard. Professor Osamu Terasaki at Stockholm University is the world's leading expert in the study of surface structure in framework materials by electron microscopy. Both the UMIST and RI groups have long-standing collaborations with Professor Terasaki. Similarly, in terms of speciation during crystal growth monitored by NMR methods, Professor Francis Taulelle from Strasbourg is currently the world's leading expert. Consequently, the purpose of this proposal is both to bring to bear the strength of our combined expertise and to effect a knowledge transfer from these leading groups to the UK for the future. This should enable the UK to consolidate its pre-eminence in the area of crystal growth of open-framework materials. Finally, one of the most exciting new developments is the potential of mass spectrometry to monitor speciation in-situ during crystal growth. This has been demonstrated by Schüth in Mülheim (Bussian et al., 2000). At UMIST, we are particularly well-equipped through the Michael Barber Centre for Mass Spectrometry (Gaskell) to integrate such measurements into these studies.

WIDE-RANGING BENEFITS

The far-reaching consequences of this work can be illustrated with a prediction. Many open-framework inorganic materials are synthesized using expensive organic templates. The full role of these templates is at present unclear, however, if the primary role during crystal growth is to promote surface nucleation, then the concentration might be reduced to a level where nucleation still occurs and a less expensive space filler may be employed to continue growth of new crystalline layers. This is a single predictive example - but one might also envisage novel controlled intergrowth structures; defect free structures; controlled crystal habit; crystals embedded within crystals; massive crystals; single crystal films; epitaxial growth of materials with complementary properties. Such complexity eludes the community at present because a far better understanding of the crystal growth process is required".

From an EPSRC chemistry grant awarded to Prof. Anderson M. William (University of Manchester Institute of Science and Technology) and co-applicants (1998).

The background section shown in Example 26.1 begins with the purpose of the project, followed by a brief description of the approach, a detailed history (equivalence of literature review) of the field and the contributions of UK scientists to the field and then ended with the benefits. The contributions of UK scientists contain a mixture of those from other experts and those from the applicants in line with Section 26.2. Another strong feature of this background section is that it can be understood by both experts and nonexperts.

FURTHER READING

Goldbort, R. 2008. *Writing for science*: Yale University Press.
- Discusses the background and literature review section of a research grant proposal in light of an NSF grant application.

REFERENCES

Agger, J.R., N. Pervaiz, A.K. Cheetham, and M.W. Anderson. 1998. Crystallization in zeolite A studied by atomic force microscopy. *J. Am. Chem. Soc.* 120 (41):10754–10759.

Agger, J.R., N. Hanif, C.S. Cundy, A.P. Wade, S. Dennison, P.A. Rawlinson, and M.W. Anderson. 2003. Silicalite crystal growth investigated by atomic force microscopy. *J. Am. Chem. Soc.* 125 (3):830–839.

Anderson, M.W., J.R. Agger, J.T. Thornton, and N. Forsyth. 1996. Crystal growth in zeolite Y revealed by atomic force microscopy. *Chem. Int. Ed. Engl.* 35 (11):1210–1213.

Anderson, M.W., J.R. Agger, N. Hanif, and O. Terasaki. 2001. Growth models in microporous materials. *Microporous Mesoporous Mater.* 48 (1):1–9.

Bazzana, S., S. Dumrul, J. Warzywoda, L. Hsiao, L. Klass, M. Knapp, J.A. Rains, E.M. Stein, M.J. Sullivan, C.M. West, J.Y. Woo, and A. Sacco. 2002. Observations of layer growth in synthetic zeolites by field emission scanning electron microscopy. *Stud. Surf. Sci. Catal.* 142:117–124.

Binnig, G., H. Rohrer, C. Gerber, and E. Weibel. 1982. Surface studies by scanning tunneling microscopy. *Phys. Rev. Lett.* 49 (1):57–61.

Binnig, G., C.F. Quate, and C. Gerber. 1986. Atomic force microscope. *Phys. Rev. Lett.* 56 (9):930–933.

Burton, W.K., N. Cabrera, F.C. Frank, and N.F. Mott. 1951. The growth of crystals and the equilibrium structure of their surfaces. *Phil. Trans. R. Soc.* A 243 (866):299–358.

Bussian, P., F. Sobott, B. Brutschy, W. Schrader, and F. Schüth. 2000. Speciation in solution: Silicate oligomers in aqueous solutions detected by mass spectrometry. *Angew. Chem. Int. Ed.* 39 (21):3901–3905.

Dumrul, S., S. Bazzana, J. Warzywoda, R.R. Biederman, and A. Sacco. 2002. Imaging of crystal growth-induced fine surface features in zeolite A by atomic force microscopy. *Microporous Mesoporous Mater.* 54 (1):79–88

Gibbs, J.W. 1928. *Collected Works*: Longman, New York.

McPherson, A., A.J. Malkin, and Y.G. Kuznetsov. 2000. Atomic force microscopy in the study of macromolecular crystal growth. *Ann. Rev. Biophys. Biomol. Struct.* 29 (1):361–410.

Singh, R., J. Doolittle, M.A. George, and P.K. Dutta. 2002. Novel surface structure of microporous Faujasitic-like zincophosphate crystals grown via reverse micelles. *Langmuir* 18 (21):8193–8197.

Wiseman, J.T, K. Alavi, and R.J. Milner. 2013. Grant writing 101. *Clin. Colon Rectal Surg.* 26 (04):228–231.

Wulff, G. 1901. Zeitschrift für Kristallographie. *Kristallogr. Kristallgeom.* 34:949.

Zaremba, C.M., A.M. Belcher, M. Fritz, Y. Li, S. Mann, P.K. Hansma, D.E. Morse, J.S. Speck, and G.D. Stucky. 1996. Critical transitions in the biofabrication of abalone shells and flat pearls. *Chem. Mater.* 8 (3):679–690.

Zlowodzki, M., A. Jönsson, P.J. Kregor, and M. Bhandari. 2007. How to write a grant proposal. *Indian J. Orthop..* 41 (1):23–26.

27 Methodology of a Research Grant Proposal

27.1 FUNCTION OF THE METHODOLOGY SECTION OF A RESEARCH GRANT PROPOSAL

The methodology section is the "heart" or "blue print" (Marshall, 2013) of a research grant proposal. The primary function of the methodology section is to provide details of the procedures, equipment and techniques that will be used to achieve the aim and objectives of the proposed research. For many funders, the methodology section makes up 50% of the overall proposal, justifying its centrality in the grant proposal (Chung and Shauver, 2008). This is where majority of the errors occur in the grant proposal (Chung and Shauver, 2008). Therefore, this section needs to be comprehensive, unambiguous and error-free to allow reviewers judge the feasibility and potential success of the proposed research (Marshall, 2013). Many grant applications are unsuccessful on the basis of this section; therefore, it is imperative to give it a special priority. Choose an appropriate method for each experiment as a wrong methodology will count against the success of the research and get the application rejected (Denscombe, 2012). Support the methodology with references from trusted sources if possible so that reviewers will be confident about the project outcomes (Ohtake, 2000, Al-Shukaili and Al-Maniri, 2017).

27.2 WRITING THE METHODOLOGY SECTION OF A RESEARCH GRANT PROPOSAL

This section builds on the objectives of the proposed research:

- Describe how each objective will be accomplished, stating possible outcomes. The description of methods should be in a logical and sequential manner of the proposed experiments (Marshall, 2013) and should include any (i) limitations, (ii) mitigations for limitations or potential problems and (iii) reasons why a particular method is chosen over others.
- Although it is good to be detailed, avoid too much details, especially for common and minor measurements (Al-Shukaili and Al-Maniri, 2017) like pH, conductivity and mass.
- Reference all literature sources related to the methods, stating any modifications as the case may be (Ohtake, 2000).
- If necessary, use diagrams and/or flow charts, especially for long procedures involving step-by-step approaches (Keshavan, 2013).

DOI: 10.1201/9781003186748-27

Methodology

- **Objective 1** – Restate verbatim
- **Experiment 1** – State the title
- **Method** – State how the experiment will be done
- **Rationale** – State why the experiment is necessary
- State anticipated results and interpretation
- State potential problems, pitfalls and alternative strategies

FIGURE 27.1   Possible structure of the methodology section of a research grant proposal.

- If the methodology involves the use of animal or human tissues or cells, it is likely that an ethical approval will be needed. Be sure to get this approval and include it in the methodology (Keshavan, 2013).
- The method for each objective can be structured according to Figure 27.1, as exemplified in Grant Proposals I and II.

FURTHER READING

Denscombe, M. 2012. *Research proposals: A practical guide:* McGraw-Hill Education (UK).
- Contains practical suggestions for writing the methodology section of a research grant proposal.

Goldbort, R. 2008. *Writing for science*: Yale University Press, New Haven.
- Contains useful suggestions for writing the methodology section of a research grant proposal.

REFERENCES

Al-Shukaili, A., and A. Al-Maniri. 2017. Writing a research proposal to the Research Council of Oman. *Oman Med. J.* 32 (3):180–188.

Chung, K.C., and M.J. Shauver. 2008. Fundamental principles of writing a successful grant proposal. *J. Hand Surg. Am.* 33 (4):566–572.

Denscombe, M. 2012. *Research proposals: A practical guide:* McGraw-Hill Education, Berkshire, UK.

Keshavan, M.S. 2013. How to write a grant and get it funded. *Asian J. Psychiatr.* 6 (1):78–79.

Marshall, L.S. 2013. Research commentary: Grant writing: Part II grant application/proposal components. *J. Radiol. Nurs.* 32 (1):48–51.

Ohtake, P.J. 2000. Research corner: Grant writing: Toward preparing a successful application. *Cardiopulm. Phys. Ther. J.* 11 (2):69–73.

28 Budgeting

28.1 FUNCTION OF THE BUDGET SECTION OF A RESEARCH GRANT PROPOSAL

The function of the budget section of a research grant proposal is to provide a detailed breakdown of the research cost. This serves as the financial plan of the proposed project. Funders normally stipulate the maximum amount of money they will give per research project, and it is advisable to make your budget within the limit of this amount. Exceeding the budget limit of the funder will automatically render your application unsuccessful. Likewise, an inflated budget, even if it is within the funding limit of the funder, and a low budget will render the application unsuccessful as reviewers will take it that the applicants do not know what they are doing (Kaplan, 2012).

28.2 WRITING THE BUDGET SECTION OF A RESEARCH GRANT PROPOSAL

The budget section of research grant proposals can be broadly divided into two subsections: (i) direct and (ii) indirect costs (Patil, 2019). Irrespective of the cost, be sure:

- That the funder has provision for it before including it on the budget. Your application will be unsuccessful if your budget contains costs that the funder does not have provision for. Ask for clarification if the funder's guideline is unclear about certain costs by emailing, telephoning or faxing the contact person(s).
- That your budget is realistic (Keshavan, 2013) and clearly justified (Patil, 2019), *e.g.*, the cost of an equipment can be justified by stating the importance of the equipment for the project. The cost of each item on the budget should be justified with the frequency of usage clearly stated.
- Not to exhaust the stipulated amount, unless it is necessary, as funders and reviewers often question the justification for this.

28.2.1 DIRECT COSTS

Direct costs (Example 28.1) are expenses specifically related to carrying out the research project, and these include expenses toward materials (equipment, consumables, *etc.*) and personnel (salaries, travels, fringe benefits, *etc.*). These expenses may be further subdivided into those that occur frequently (recurring) and those that occur less frequently (non-recurring) during the project duration.

DOI: 10.1201/9781003186748-28

a. *Materials cost*: Refers to expenses toward purchase of equipment and consumables. Expenses toward equipment are normally non-recurring while those toward consumables may be recurring.

b. *Personnel cost*: Refers to expenses of all the people involved in the research project. This may include salaries and travels (*e.g.*, conference, meeting or workshop attendance). Expenses toward salaries fall under recurring costs.

28.2.2 INDIRECT COSTS

Indirect or administrative costs (Example 28.1) are not specifically related to the project cost, but they are needed to successfully run the project and they include overhead costs (Patil, 2019). Overhead costs are expenses toward institutional facilities like laboratory space, library access, water and electricity supplies and are paid directly to the institution.

Example 28.1 *The Budget Section of a Research Grant Proposal*

The budget section of a successful EPSRC chemistry grant proposal (Appendix III). The grant was awarded to Prof. Anderson M. William (University of Manchester Institute of Science and Technology, UK) and others in 1998.

	Total/£
Staff	268,065
Travel and subsistence	33,720
Consumables	59,987
Exceptional items	34,440
Equipment	14,837
Large capital	194,450
Sub-total	**605,499**
Indirect costs	123,310
Total	**728,809**

Activity 28.1 *Identifying Recurring and Non-Recurring Costs in a Research Grant Proposal*

Identify possible recurring and non-recurring costs in Example 28.1.

28.3 BUDGET JUSTIFICATION AND BUDGET SUMMARY OF A RESEARCH GRANT PROPOSAL

The *budget justification* or *narrative* section of a grant proposal explains the importance of an item on the proposed project (Patil, 2019). In order words, it should make a strong case for why the funder should spend money on the item. For instance, if the

budget contains money for purchasing an optical microscope, the budget justification section gives reasons why the optical microscope is necessary for the proposed
project. The reasons could be to (i) view microorganisms, (ii) view sample specimens
or (iii) follow morphological changes in a sample specimen. The best format for the
budget justification section is to follow the order of items on the budget (Patil, 2019),
either in a tabular or textual form (Example 28.2).

Example 28.2 *The Budget Justification Section of a Research Grant Proposal*

JUSTIFICATION OF RESOURCE (GRANT PROPOSAL I, APPENDIX III)

"The total amount of funding requested is £968,000 which will support the
research of over 19 researchers and academics in a truly unified project. We
believe that by bringing together leading researchers in the field of microporous
materials to tackle a major scientific problem, our proposed network represents
outstanding value for money (less than £50k per researcher) and is likely to make
far more impact than individual researchers working in isolation.

The main cost of the project is the personnel cost for a large team with interrelated expertise. This will involve 7 PhDs, 1.5 PDRAs and technical support. Staff
numbers and seniority have been carefully chosen to ensure the team is able to
carry out a wide range of activities to the necessary level of competence and
within the timeframe of the project. In this respect, one PDRA and PhD student
will work on AFM and synthesis with the PDRA concentrating more on in-situ
AFM development and the student on ex-situ measurements. One PhD student
will work on solution phase speciation using NMR and mass spectrometry – this
student will also perform in-situ XRD measurements at station 16.4, Daresbury (for
which time will be requested separately) to define the crystallization co-ordinate
for their speciation studies. One PhD student and 0.5 PDRA will concentrate on
electron microscopy measurements of surface structure. This is a complex technique requiring the full dedication of separate personnel. One PhD student will
carry out computer simulation of crystal morphology and topography in order
to extract kinetic and thermodynamic information from the experimental data.
Finally, three PhD students will work on classical and *ab initio* modeling of surface
structure, binding energies and rates of attachment of growth units and cation
templating studies. The experimental synthetic work will be supported by 20% of
an experienced technician. This is the minimum number of personnel to perform
this complex range of tasks.

This proposal relies on importing some knowledge from overseas in terms of a)
high-resolution TEM and ultra-high-resolution SEM from the groups of O. Terasaki
and V. Alfredsson and b) in-situ NMR of crystallization from F. Taulelle. The idea
is to establish a knowledge base that can be retained in this country within the
UCMM after the end of this project. This necessitates two of the PhD students
carrying out 1/3 of their studies abroad for which travelling and a modest extra
subsistence is requested. Travelling is also requested to hold 6 monthly project
meetings of all members in Manchester. Attendance at one international conference is requested for the team as well as annual participation by the students at
the British Zeolite Association (BZA) Conference. Equipment for the project comprises: an AFM with high-resolution optical microscope, for locating micron-sized

crystals and variable temperature capability for in-situ studies; AFM probes; a Cr sputterer for the high-resolution SEM studies at UMIST; we have requested additional monies to provide 1 Linux-based high-performance workstation for each simulation PhD student. These machines will be dedicated computer resources for undertaking the numerically intensive simulation work. Running and access costs are also requested for NMR, MS and TEM/SEM at UMIST".

From an EPSRC chemistry grant awarded to Prof. Anderson M. William (University of Manchester Institute of Science and Technology) and co-applicants (1998)

Contrarily, the *budget summary* section gives an item-by-item summary of the yearly expenditure of the total budget for the entire duration of the project, *e.g.*, 2, 3, 4 years, *etc.* Some funders give provision for the budget, budget justification and summary in one table while others split them into different sections. Be sure to follow which format applies to your application. The budget summary of one of the items (travels and subsistence) in Example 28.1 is given in Example 28.3. Budget summaries for the other items in Example 28.1 can be found in Grant Proposal I (Appendix III).

Example 28.3 *The Budget Summary of Travels and Subsistence Given in the Grant Proposal Budget Shown in Example 28.1*

Destination and Purpose	Total/£
(i) Within UK	
Four PhDs to annual BZA meeting	2000
Six staff to one BZA conference	3000
Outside UK -four PhDs and six staff to IZA meeting	15,000
(ii) Outside UK	
6× Sweden-Manchester with four nights' accommodation for Terasaki and Alfrdesson (six monthly meetings)	7920
6× France-Manchester with four nights' accommodation for Taulelle (six monthly meetings)	3360
2× Sweden-Manchester with four nights' accommodation for PhD microscopy (two of the six monthly meetings)	1320
2× France-Manchester with four nights' accommodation for PhD NMR MS XRD (two of the six monthly meetings)	1120
TOTAL	33,720
Source: From an EPSRC chemistry grant awarded to Prof. Anderson M. William (University of Manchester Institute of Science and Technology) and co-applicants (1998).	

Activity 28.2 *Preparing a Budget, Budget Justification and Summary for a Project*

Prepare a budget, budget justification and budget summary for a future experiment in your laboratory. Are there recurring and non-recurring costs in the budget?

FURTHER READING

Crawley, G.M., and E. O'Sullivan. 2015. *The grant writer's handbook: How to write a research proposal and succeed.* World Scientific.
- Contains useful suggestions for writing the budget and other sections of a research grant proposal.

Goldbort, R. 2008. *Writing for science*: Yale University Press, Binghamton, New York.
- Discusses the budget section of research grant proposals and gives useful suggestions for writing the section.

Kanji, S. 2015. Turning your research idea into a proposal worth funding. *Can. J. Hosp. Pharm.* 68 (6):458–464.
- Contains tips for writing the budget section of a research grant proposal for first timers.

REFERENCES

Kaplan, K. 2012. Funding: Got to get a grant. *Nature* 482 (7385):429–431.

Keshavan, M.S. 2013. How to write a grant and get it funded. *Asian J. Psychiatr.* 6 (1):78–79.

Patil, S.G. 2019. How to plan and write a budget for research grant proposal? *J. Ayurveda Integr Med.* 10 (2):139–142.

29 Timeline

29.1 FUNCTION OF THE TIMELINE SECTION OF A RESEARCH GRANT PROPOSAL

The timeline, also known as the work plan, serves as the "time management plan" or "time table" of the proposed project. The timeline typically shows the tentative start and end dates of the project as well as the duration of various project tasks. Some funders stipulate strict start and end dates; therefore, it is advisable to prepare your timeline in line with these dates.

29.2 WRITING THE TIMELINE SECTION OF A RESEARCH GRANT PROPOSAL

The timeline should be feasible (Goldbort, 2008) and should show the start and end dates of all project tasks. The timeline can be prepared in a tabular (Example 29.1), graphical (*e.g.*, the Gantt chart, Example 29.2) or textual (Example 29.3) form. A Gantt chart, named after its inventor (Henry Gantt), is a form of bar chart that shows a project schedule. The chart shows the tasks on the vertical axis and the time allotted to them on the horizontal axis, with the widths of the bars representing the duration of each task. The Gantt chart can also be used to show the relationship between the tasks. The timeline in Grant Proposal I is presented as a Gantt chart. Some funders specify their acceptable timeline format (tabular, Gantt chart or textual), be sure to adhere to this specification. In the absence of any specifications, you are welcome to adopt any format, but ensure that the time is unambiguous. More importantly, the timeline should encapsulate each objective or specific aim. For illustration, the timeline of Grant Proposal V (Appendix III) is shown in different formats in Examples 29.1–29.3. Remember that irrespective of the format, the guiding principles are *clarity* and *capturing* of each objective (specific aim, key tasks or activities) on the timeline.

DOI: 10.1201/9781003186748-29

Example 29.1 *Tabular Form of a Grant Proposal Timeline, the gray shade represents time range*

S/No	Task	June–July	August	September
1	Test strain gauges in the lab (mechanical and electrical tests)	▓▓		
2	Further investigate device adhesion issues	▓▓	▓▓	
3	Calibrate strain gauges		▓▓	
4	Test bone samples (*in vitro*)			▓▓

Example 29.2 *Gantt Chart Form of a Grant Proposal Timeline, the gray shade represents time range*

Tasks	June–July	August	September
Test strain gauges in the lab (mechanical and electrical tests)	▓▓		
Further investigate device adhesion issues	▓▓	▓▓	
Calibrate strain gauges		▓▓	
Test bone samples (*in vitro*)			▓▓

Example 29.3 *Textual Form of a Grant Proposal Timeline*

June–July	Test strain gauges in the lab (mechanical and electrical tests)
August	Further investigate device adhesion issues
August	Calibrate strain gauges
September	Test bone samples (*in vitro*)

Activity 29.1 *Preparing a Grant Proposal Timeline*

Using any of the formats described here, prepare a timeline for a future experiment in your laboratory.

REFERENCE

Goldbort, R. 2008. *Writing for science*. Yale University Press: Binghamton, New York.

30 Summary of Research Grant Components

Component	Main Function	Verb Tense	Contents
Title			
One-sentence summary of the proposal's essence	Tells the reader what the proposed research is all about	Present tense	Keywords or phrases (little or no jargon)
Details of Applicant(s)			
Name, qualification and addresses of applicants	Tells the reader who the applicants are, their qualifications and their addresses		
Abstract			
The table of contents or synopsis of the proposal	Gives the reader a bird's eye-view of the proposed project	Future tense, referring to what will be done	Brief background, aim of work, methods (how the work will be done), preliminary data (if available), significance and cost of project
Summary			
Summary of the proposed project	Gives a summary of the proposed project	Future tense	All the components of the proposal, except references
Introduction			
Framework of the proposal	Gives the reader the framework of the proposed project	Mixture of present and past tenses as well as future tense	Background, statement of problem, research gap, aim and objectives of project, significance of project and competency claim

DOI: 10.1201/9781003186748-30

Component	Main Function	Verb Tense	Contents
Background and Literature Review			
Background information about the work and its relationship with the literature	Sets the scene and shows how the proposed work is related to earlier ones	Present tense (established knowledge), present perfect tense (previous research) and past tense (unestablished knowledge)	Background information, literature review of previous researches and justification of the work
Methodology			
How the work will be carried out	Gives a vivid description of how the proposed project will be carried out	Future tense	Detailed description of materials to be used and experimental procedures
Budget			
The cost of carrying out the proposed project	Detailed breakdown of the cost of carrying out the project		Direct and indirect cost as well as budget justification
Timeline			
The amount of time needed to complete the research	Provides a detailed time management plan of the project		Start and end dates
References			
List of supportive or appropriate literatures	Gives a list of relevant literatures related to the proposed work		List of all the relevant literatures related to the proposed project

31 Research Grant Proposal Review Process

31.1 RESEARCH GRANT PROPOSAL REVIEW

Submitted research grant proposals are subjected to expert review. Some funders give applicants the opportunity to propose potential reviewers (usually three or so) of their proposal and also name those they would not like to review their proposal. Count this as an opportunity to influence the review in your favor and choose reviewers wisely. My best advice here is to choose people who would generally support your proposal. The funder may go with one or none of your suggestions. However, the suggestions give them an idea of who to and who not to contact for expert review. For most funders, submitted proposals undergo a three-stage review process, as illustrated in Figure 31.1. In the first stage, the funder's desk manager screens the proposals based on pre-set eligibility criteria as well as writing guidelines and separates them into two categories, namely "to be reviewed" and "not to be reviewed". Proposals that are prepared in line with the funder's guidelines and the applicants also meet the eligibility criteria form the first category. Contrarily, proposals that did not follow the funder's guidelines and/or the applicants did not meet the eligibility criteria form the second category. While the "not to be reviewed" proposals are not considered further, the "to be reviewed" proposals are sent to experts for review.

The reviewers scrutinize and score the proposals in terms of (Tseng, 2011, Ohtake, 2000):

a. *Originality or innovativeness of the research idea*: Funded ideas are normally original and innovative.
b. *Significance of the proposed research and its potential impact*: Funded research projects are expected to be impactful.
c. *The workability and the validity of the proposed methodology*: The proposed methodologies are expected to be adequate for obtaining the desired outcomes of the project.
d. *The feasibility of the scope of the proposed project*: Are the applicants biting more than they can chew? In other words, can the project be completed within the proposed timeframe? Overambitious projects, which cannot be completed within the proposed timeframe are normally not funded.
e. *The background information supporting the proposed project*: Does the background information provided support the project or not? Reviewers expect the background information provided to support the project's importance and impact.
f. *The expertise and experience of the research team, especially the principal investigator*: Reviewers expect the research team to be experts who have experience in the proposed area of research.
g. *The feasibility of the budget*: Proposals with inflated or underestimated budgets never gets funded.

DOI: 10.1201/9781003186748-31

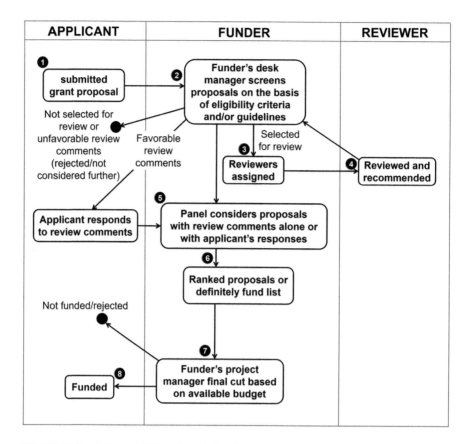

FIGURE 31.1 Schematic illustration of the life cycle of a research grant proposal, from submission to funding. Submitted proposals are screened by the funder's desk manager on the basis of applicant eligibility criteria and writing guidelines. Proposals that fall short of these are not processed further while those that measure up are sent to experts for review. Reviewers revert back to the desk manager with comments and recommendations. Proposals with unfavorable comments are not processed further while those with favorable comments are either sent to applicants for responses before consideration at a panel or considered by the panel without applicant responses. The panel ranks the proposals from the best to the least or creates a "definitely fund" list. The ranked proposals or the "definitely fund" list is sent to the project manager who decides the number of proposals to fund *vis-à-vis* the available budget for that round or year.

    h. *Adequacy of research facilities*: Reviewers check that the research team has adequate research space and equipment to carry out the proposed research.

Additionally, some funders also ask reviewers to state whether or not they should fund the proposed research. Generally, proposals with unfavorable review comments are not considered further while those with favorable review comments move to the next review stage. Some funders like EPSRC give applicants the opportunity to respond to review comments. This gives applicants the chance to clarify issues that may have been misunderstood by reviewers. Typical research grant proposal review comments can be found in Grant Proposals II (page 10) and VIII (page 84). The review comments in Grant Proposal II have been addressed in Grant Review

Comments I (Appendix III), which can serve as a guide for responding to review comments of research grant proposals. In the second stage, the review comments alone, or with applicants' responses (for EPRSC), and the proposals are read by a panel. Such panels would normally consist of both experts and nonexperts of the subject area. The panel may either rank the proposals, starting with the best to the least, or categorize them into three groups, namely "definitely fund", "probably fund" and "do not fund". The "definitely fund" proposals are those with favorable review comments and high scores. Comparatively, the "probably fund" proposals may have favorable review comments, but moderate scores while the "do not fund" proposals have both unfavorable review comments and low scores. Finally, the funder's program manager takes the ranked or "definitely fund" proposal list and the associated project cost and matches the cost with the budget available for that round or year. The funder then draws a limiting line on the list in accordance with the available budget.

Activity 31.1 *Comparing a Scientific Research Paper and a Research Grant Proposal*

Findings from the proposed research of Grant Proposal VIII (Appendix III) have been published by the researchers (Tyler et al., 2021), thus allowing direct comparison between a research grant proposal and a scientific research paper. Compare the grant proposal with the published paper in terms of structure and language style. Pay particular attention to their titles, abstracts, introductions, methodology and references.

31.2   FINAL OUTCOMES

There are only two possible final outcomes from research proposal review: funded or rejected. Congratulations! If your research proposal is accepted and funded. It is time to assemble the human and material resources together to execute the project. Keep the funders informed about the progress of the project as well as any changes to the project. However, if your proposal is rejected, all hope is not lost – you can try again; even proposals from very experienced researchers often get rejected. Some funders provide review comments to applicants. Rather than feel frustrated, use the comments to revise your proposal and resubmit for future application rounds or even another funder. Such revised research proposals often have higher chances of being funded than their initial versions (Kanji, 2015, Wegener and Katan, 2018). Busy funders do not often return review comments, but you can call or email them with a request for the comments. Lastly, be aware that proposals are rejected not necessarily because they fall short of the points listed in Section 31.1. A proposal can also be rejected because there are insufficient funds to fund all the good proposals submitted for that round of application (Wegener and Katan, 2018). Irrespective of the reason for rejection, the golden rule for grant success is "do not give up, keep trying until you succeed".

Activity 31.2 *Comparing the Review Process for Scientific Research Papers and Research Grant Proposals*

Compare the review process for scientific research papers (Figure 19.1) with that of research grant proposals (Figure 31.1). How similar and/or different are they?

FURTHER READING

Crawley, G.M., and E. O'Sullivan. 2015. *The grant writer's handbook: How to write a research proposal and succeed.* World Scientific.
- Discusses the review process of research grant proposals and gives suggestions for responding to review comments if the funder gives the opportunity.

Ryan, B. 2004. Make your grant proposal successful. *Pharm. J.* 272:475–477.
- Contains useful suggestions for preparing a successful grant proposal and also discusses the review process of grant proposals albeit for EPSRC applications.

Peer Review Demystified (2002) by Professor Peter Scott (Appendix III) of the University of Warwick, United Kingdom, gives comparing insight to grant proposal review albeit for EPSRC. The life cycle of a research grant proposal from the final draft to funding, or otherwise, is described vividly. Suggestions for responding to review comments are also given.

REFERENCES

Kanji, S. 2015. Turning your research idea into a proposal worth funding. *Can. J. Hosp. Pharm.* 68 (6):458–464.

Ohtake, P.J. 2000. Research corner: Grant writing: Toward preparing a successful application. *Cardiopulm. Phys. Ther. J.* 11 (2):69–73.

Tseng, M.-Y. 2011. The genre of research grant proposals: Towards a cognitive–pragmatic analysis. *J. Pragmat.* 43 (8):2254–2268.

Tyler, K., S. Geilman, D.M. Bell, N. Taylor, S.C. Honeycutt, P.I. Garrett, T.M. Hillhouse, and T.M. Covey. 2021. Acyl peptide enzyme hydrolase (APEH) activity is inhibited by lipid metabolites and peroxidation products. *Chem.-Biol. Interact.* 348:109639.

Wegener, S., and M. Katan. 2018. Getting the first grant. *Stroke* 49 (1):e7–e9.

32 Book Proposals

32.1 TYPES OF BOOKS

Books in chemistry fall into research books (*e.g.*, proceeding volumes, monographs and handbooks; see Section 1.2) and textbooks. Research books are specialized professional books written for scholars (*i.e.*, postgraduate students, early-career and established researchers). Research books do not contain worked examples and may also not contain end-of-chapter questions. By contrast, textbooks are written for lower-level students like college and undergraduate students. As a result, textbooks are generally packed with detailed information about the subject area, worked examples and end-of-chapter questions to enable the students learn fundamental concepts of the subject area and also assess their level of understanding, with or without the guide of an instructor. Such textbooks are sometimes accompanied by study videos to aid self-learning and cement key concepts in students and an instructor manual to guide instructors. Textbooks are expected to satisfy the instructional needs of the instructors before they can be adopted for teaching.

Activity 32.1 *Differences between Research Books and Textbooks*

In tabular form, list the differences between research books and textbooks. Also, classify books in your collection or library as research books and textbooks.

32.2 CHOOSING A PUBLISHER

Choosing a publisher depends on the type of book you are aiming to write. Some publishers prefer to handle research books while others rather prefer to handle textbooks. Some publishers are into both research books and textbooks. Checking publishers' websites or contacting them (*via* email, phone or fax) with questions regarding the types of books they are into can help you decide the publisher to go for. Publishers like the Oxford University Press, Cambridge University Press, Elsevier, Wiley, Taylor and Francis and Springer are into both research books and textbooks while publishers like the Royal Society of Chemistry and the American Chemical Society are rather happier to publish research books than textbooks.

32.3 WRITING A BOOK PROPOSAL

After deciding your publisher, your next task is writing a proposal for your book idea. A book proposal gives the publisher an overview of the proposed book. Publishers generally have templates for their book proposals which can be downloaded from their websites or sent to you by a commissioning editor upon sharing your book idea with the company. Such proposal templates generally ask the following information:

DOI: 10.1201/9781003186748-32

    a. Book type and title.
    b. Author(s) information.
    c. Brief author biography.
    d. Background and aim of the book.
    e. Book description.
    f. Unique features of the book and potential price.
    g. List of possible competitors and their prices.
    h. Table of contents of the book.
    i. Sample chapter.
    j. Tentative specifications of the book.
    k. The software packages you are going to use in preparing a draft copy of the book.
    l. Timeframe for submitting a draft copy of the book.
    m. Marketing opportunities/audience for the book.
    n. List of potential reviewers of the book.
    o. Funding information.

The aim of these information is to give the publisher a clear picture of the proposed book as well as the author. The next subsections describe how to provide the requested information in your book proposal form.

32.3.1 BOOK TYPE AND TITLE

State whether the proposed book is a research book or a textbook. In providing a working or a tentative title of the book, use keywords readers are likely to use in an online search for information on the subject area. The book title should:

- Accurately describe the content of the book.
- Not mislead readers to other subject areas.
- Not be too long ($\leq$50 characters if possible).

32.3.2 AUTHOR(S) INFORMATION

Provide a complete list of all the authors or editors (for edited research books) along with their contact information (*e.g.*, affiliations, email addresses, faxes, call lines and ORCID iDs).

32.3.3 BRIEF AUTHOR BIOGRAPHY

Provide a complete biography of all the authors or editors (for edited research books) involved in the book project. This should include their qualifications and experience in the subject area of the proposed book. The biography should also state why you or these authors are the most suitable for writing a book on the subject area, *e.g.*, it could either be that you are a leading authority or these authors are leading authorities in the subject area of the proposed book. It is also helpful to supply the curriculum vitae of the authors (even if the publisher does not ask).

32.3.4 BACKGROUND AND AIM OF THE BOOK

Think of the following questions while responding to this section:

- What is the significance of the proposed book?
- What is the aim of writing the book?
- Why is now the right time to publish a book on the topic?

Significance: Explain the significance of the subject area and the topic of the proposed book. This can be supported with the frequency of the proposed topic in the news and the social media, scientific conferences, scholarly journal publications and citations – statistics can be used to drive the point home. If possible, explain how the topic overlaps with other subject areas, examining various views of authors in these areas and drawing parallels between the views that will be presented in the proposed book.

Aim: State clearly the purpose of the proposed book, which could be to (i) augment existing books with new developments in the subject area, (ii) provide a reference book for new comers to the subject area, (iii) fill a book vacuum for the subject area, *etc.*

Timing: Why does the subject area need the book now? It could be that the proposed book will accelerate the development of the subject area or help new comers to the subject area overcome experimental hurdles, *etc.*

32.3.5 BOOK DESCRIPTION

The book description says what the book is all about and should be directed mainly to the target audience, but should make sense to a layman too. In writing the book description, bear in mind that the critical question here is "why should someone (audience) buy the proposed book?" Put it differently, "what will someone (audience) lose by not buying the proposed book?" It could be that the proposed book will (i) provide a detailed information about the subject area compared with the existing ones, (ii) update the reader with new advances in the subject area, *etc.* Therefore, someone will lose these by not buying the proposed book. The first sentence in your book description should say "this book will provide/offer …". Next, list the main topics that will be covered, stating why they are important to the audience and what you have set out to do in the book. End this section with a brief description of the target audience of the book.

32.3.6 UNIQUE FEATURES AND POTENTIAL PRICE

What makes the proposed book unique from other books? Maybe the proposed book (i) provides a more detailed review of the subject area, (ii) provides more illustrations of practical applications of the subject area and (iii) covers evolving and special topics on the subject area than other books. All of these constitute the unique selling features of the proposed book. The next question is "how much do you think a book with these unique and wonderful features should cost?" It is always helpful to check the prices of similar books on the subject area online and then infer a price range that is fare for the proposed book.

32.3.7 COMPETING BOOKS AND THEIR PRICES

This is where authors often lie, saying "there are no competing titles for the proposed book". To the publisher, saying this means you have not thoroughly researched the market for the proposed book or there is no market for the proposed book; thus, there is no need to pursue the book project further. Therefore, be sincere in your response, even if there are strong competing titles out there. In fact, if the proposed book is well-researched, well-written and well-timed, it can compete favorably with well-established competing titles or even outdo them. Your ability to demonstrate the position of the proposed book in the market will significantly increase the likelihood of your proposal to sail through. An online search, book stores, libraries, discussion with colleagues and the target audience can give you clues to competing titles (*i.e.*, research books, textbooks and review articles) on the subject area of the proposed book as well as their prices. Using a table, make a comprehensive list of these titles along with their authors, publishers, prices, strong and weak points. Finally, describe how the proposed book is unique from these titles and how it takes care of the weak points identified in the competing titles. This will make your proposal stronger and more likely to succeed.

32.3.8 TABLE OF CONTENTS AND SAMPLE CHAPTER

The table of contents gives the publisher an idea of the structure of the proposed book. Taking into cognizant of the subject area, the target audience and the position of the proposed book in the market, the chapter titles should be clear and succinct. Each chapter should be accompanied by a synopsis and a brief abstract to enable reviewers give helpful feedback on both the content and the structure of the book. A sample chapter is not mandatory, but when available, it helps reviewers to assess the writing style of the author(s), grammar proficiency and potential of the book proposal to successfully executed.

32.3.9 BOOK SPECIFICATIONS

Publishers like to know approximately the number of pages, figures (colored and monochrome), graphs, tables, mathematical and chemical equations in the proposed book to enable them make appropriate budget for its production. Be sure that your approximation of these is as fare as possible.

32.3.10 SOFTWARE FOR PREPARING DRAFT COPY

Publishers may prescribe the software that must be used in preparing a draft copy of the proposed book or ask authors to tell them what software are available to them for the preparation of the draft copy (*i.e.*, text, figures, tables, graphs and sketches). This helps them check software compatibility issues with their publishing machines and style. Microsoft Office is a common software for compiling text and preparing tables, graphs and sketches; thus, it is widely accepted by publishers so feel free to use it. Alternative word processing software like LaTeX can also be used if the publisher authorizes it.

32.3.11 TIMEFRAME FOR SUBMITTING A DRAFT OF THE PROPOSED BOOK

Publishers like to know how long it will take until a draft copy of the proposed book will be available for review and subsequently publication. Writing a book takes time (at least a year), especially for academics who divide their time between teaching, research, supervision, family, *etc.* This timeframe is also significantly influenced by the length and type of the proposed book as well as the wiliness of the contributors (for multi-authored books) to adhere to deadlines. Therefore, take cognizant of these factors and propose a realistic timeframe (1.5 years or so) for completing a draft copy of the book. During this timeframe, feel free to ask the publisher for an additional time if you foresee that the earlier proposed deadline cannot be met.

32.3.12 MARKETING OPPORTUNITY/AUDIENCE

Who are you expecting to buy the proposed book: researchers (early-career or established researchers and/or development scientists), students (undergraduate or postgraduate) or practitioners (*e.g.*, industrial scientists who are into a particular product) of the subject area? These are the audience for the proposed book, and they also constitute the marketing opportunity for the proposed book. Successful authors understand their audience and their needs, and they are able to satisfy these needs. Telling the publisher who the prospective audience are helps them in advertising the published book so that it reaches the desired audience.

32.3.13 POTENTIAL REVIEWERS

Publishers give authors the opportunity to suggest experts of the subject area of the proposed book who will provide useful feedback on the content and structure of the book. These experts can be deduced from their scholarly contribution on the subject area which might be available on Google Scholar, ResearchGate, LinkedIn, ORCID or Academia along with their contact details. Publishers do not necessarily use this suggestion for review purposes, but it gives them an idea of the category of people to contact for feedback on the proposed book. The publisher may also ask you to tell them the people you do not want their feedback on the proposed book. These might be people that criticize your work overtly or people that you are rather reserved about their contributions on the subject area.

32.3.14 FUNDING INFORMATION

Publishers like to know whether or not the proposed book project is going to be funded. For funded books, a transfer of copyright from the funder to the publisher is required before the publisher can publish the book.

32.4 THE REVIEW PROCESS

Submitted book proposals are first reviewed by the commissioning editor of the subject area, in conjunction with colleagues, for suitability of the book in their scope of publishing and then by external experts of the subject area. The external expert reviewers examine the structure of the book and the capability of the author(s) to write on the subject area. In their feedback to the publisher and the author(s), the reviewers (i) may suggest additional chapters for certain important topics, (ii) may re-structure the contents of the book, (iii) will state whether or not there is a market for the book and (iv) state whether or not the authors are qualified to write on the subject area. Scan the QR code in Appendix IV for a typical book proposal review comments. If these comments are favorable, the authors are commissioned to write the book. The author(s) then submit a draft copy of the book, within an agreed time-frame, for a second round of review and then publication. Contrarily, the authors are not commissioned to write the book if the review comments are unfavorable. In this case, my best suggestion to the authors is not to give up. Rather, they should consider the review comments as impactful suggestions to improve and revise their proposal. The improved and revised proposal, which now has higher chances of acceptance, can be resubmit to another publisher. The proposal I submitted to Taylor and Francis for this very book is given in Appendix IV and can be used as a practical guide or a template for a book proposal. Although two additional aspects, namely poster and oral presentations, were later introduced after the proposal was approved, the infor-mation contained in the proposal is in line with the suggestions given in this chapter.

Activity 32.2 *Analyzing a Book Proposal*

Analyze the book proposal given in Appendix IV and answer the following ques-tions: (i) What is the essence of the proposed book? (ii) What are the unique fea-tures of the proposed book compared with existing ones on the market? (iii) What value for money will readers gain by adopting or purchasing a copy of the pro-posed book?

FURTHER READING

Haynes, A. 2010. *Writing successful academic books*. Cambridge University Press.
- Contains useful suggestions for writing academic books. Suggestions are given for managing all the stages of the book project.

How to Structure Your Book Proposal (routledge.com)
- Contains useful suggestions, from commissioning editors of Taylor and Francis, for writing a book proposal. The book also contains responses to frequently asked questions.

How to write a successful book proposal (routledge.com)
- Contains useful tips, from commissioning editors of Taylor and Francis, for writing a successful book proposal.

Appendix I
General Resources on Writing

Organization of ideas

1. The Message Box – COMPASS Science Communication

Resources on grammar and word usage

2. Scientific writing booklet. Compiled by Marc E. Tischler, Department of Biochemistry and Molecular Biophysics, University of Arizona.
3. Coghill, A.M., and L.R. Garson. 2006. *The ACS style guide*: *Effective communication of scientific information*. ACS Publications.
4. Bailey, S. 2014. *Academic writing*: *A handbook for international students*. Routledge.
5. Strunk Jr, W., and E.B. White. 2007. *The elements of style illustrated*. Penguin.
6. Spector, T. 1994. Writing a scientific manuscript: Highlights for success. *J. Chem. Edu.* 71 (1):47.
7. Thomson, A.J., A.V. Martinet, and E. Draycott. 1986. *A practical English grammar*. Vol. 332. Oxford: Oxford University Press.
8. McCaskill, M.K. 1990. *Grammar, punctuation and capitalization*: *A handbook for technical writers and editors*. Vol. 7084. National Aeronautics and Space Administration, Office of Management.
9. Wiley-Blackwell House Style Guide.
10. Patience, G.S., D.C. Boffito, and P.A. Patience. 2014. Writing a scientific paper: from clutter to clarity.

Proofreading and editing

11. Cook, C. K. 1986. *Line by line*: *How to edit your own writing*: Houghton Mifflin: Boston, MA.

Quantities, units and symbols in chemistry

12. Cohen, E.R., T. Cvitaš, J.G. Frey, B. Holmström, K. Kuchitzu, R. Marquardt, I. Mills, F. Pavese, M. Quack, and J. Stohner. 2007. *Quantities, units and symbols in physical chemistry*. International Union of Pure and Applied Chemistry, The Royal Society of Chemistry.

Appendix II

Point-by-Point Responses to Peer-Review Comments for a Hypothetical Paper Based on the Suggestions Given in Section 19.3.3

Moore University, Makurdi, Nigeria
9th September 2021

The Editor/Editor-in-Chief
Journal of Chemistry Experiments (JCE)

Dear Editor/Editor-in-Chief,

**Submission of Responses to Reviewers' Comments and
Revised Version of JCE-17-31233**

We thank you and the reviewers for your time and insightful comments which have helped us improve the paper. We forward here responses to the comments raised during the review and the revised form of the paper. The paper has been thoroughly revised, in accordance with the comments. The changes we have made to the paper, in light of the comments, are highlighted in red and our specific responses are below.

REVIEWER #1

This paper describes the production of a novel kind of oil-based emulsions using Pickering stabilization. This paper is well-written and contains some useful new information, particularly about the possibility of creating novel oil-based multiple emulsions. I think it can be published after minor changes:

    a. *Introduction*: It would be useful to have more discussion about the potential applications of the emulsions.
    Response: We thank the reviewer for pointing this out. A paragraph (p. 8, paragraph 4, lines 12–22), highlighting possible applications of the emulsions has been added to the introduction as suggested.

b. *Figures 7, 10 and 15*: Standard deviations should be shown.
   Response: We completely agree with the reviewer. Thus, error bars have been added to the data points plotted in Figures 7a, 10a and 15a (p. 16, 18 and 22, respectively). However, it was impossible to do the same to the b's as that made them messy due to the large values of the standard deviations.
c. The manuscript is well-written, but it may benefit by being shortened, *e.g.*, by moving some of the data to a supplementary section (repeating the figures for the two different oils, does not add much).
   Response: We thank the reviewer for his/her kind words toward our paper. We have created a supplementary file from figures that contain similar data.

REVIEWER #2

This paper describes the formation, structure and stability of novel multiple emulsions containing two types of oils and nanoparticles. Emulsions of immiscible oils are of wide scientific interest, due to their use in applications where water must be excluded. This is innovative work that establishes the role of the particle surface chemistry in stabilizing multiple emulsions of immiscible oils with nanoparticles. In my opinion, it should be published. My only minor suggestion is that inserting higher magnification microscopy images of the multiple drops would enhance the insights provided about the emulsion microstructure and hence the impact of the article.

Response: We thank the reviewer for finding our work innovative. The optical microscope images of the emulsions in Figure 13 (p. 22) have now been improved by removing extraneous ones and enlarging the remaining ones, as suggested.

We hope the revised paper will better suite JCE, but we will be happy to undertake further revisions.

We look forward to your final decision.

Yours sincerely,

John Chemistry
(Corresponding Author)

Appendix III
Sample Grant Proposals

Sample Grant Proposal (Scan QR Code to Download)	Title, Funder and Author(s)
 Grant Proposal I	*"Crystal growth of open-framework materials"* An EPSRC chemistry grant awarded to Prof. Anderson M. William (University of Manchester Institute of Science and Technology, UK) and co-applicants (1998)
 Grant Proposal II	*"Assessing the roles of biofilm structure and mechanics in pathogenic, persistent infections"* A National Institutes of Health bioengineering grant awarded to Dr Vernita Gordon (University of Texas, Austin) and co-applicants (2017)
 Grant Review Comments I	Review comments for Grant Proposal II
 Grant Proposal III	*"Materials chemistry and thermal expansion"* A National Science Foundation grant awarded to Prof. Cora Lind-Kovacs (University of Toledo, Ohio)

(Continued)

235

(Continued)

Sample Grant Proposal
(Scan QR Code to Download)   **Title, Funder and Author(s)**

Grant Proposal IV

"Synthesis of nanowire heterojunctions for advanced nanoelectronic devices" A grant award to Prof. Philip G. Collins and Kevin Loutherback (Department of Physics and Astronomy), University of California, Irvine by the same University

Grant Proposal V

"Strain sensitive array for the study of bone surface mechanics" A grant award to Prof. William Tang (Samuel School of Engineering) and Prof. Joyce H, Keyak (Radiological Sciences School of Medicine), University of California, Irvine by the same University

Grant Proposal VI

"Contribution to the science behind alternative energy utilization: Developing the fundamental chemistry behind solar energy conversion using Ni, Pd, Pt metal complexes and small molecule activation" A grant awarded to Prof. Alan F. Heyduk (Department of Chemistry), University of California, Irvine by the same University

Grant Proposal VII

"Precipitation of carbonates by viral lysis of cyanobacteria in GSL" A grant awarded to Prof. Frantz Carie and Santana Darian (Department of Earth and Environmental Sciences), Weber State University by the same University

Grant Proposal VIII

"Investigating the inhibition of APEH in disease" A grant awarded to Prof. Tracy Covey, Deborah Belnap and Shelby Geilmann (Department of Chemistry and Biochemistry), Weber State University by the same University

Appendix IV
Sample Book Proposal

CRC PRESS

BOOK PUBLISHING PROPOSAL

Please answer the following questions as completely as possible. The information provided in this document will be used for editorial review and may be shared with peers in your field in consideration of contract approval.

Please send your completed form to barbara.knott@taylorandfrancis.com.

(Spaces will expand as needed to accommodate longer answers)

1. **Title of your book:** A Practical Guide to Scientific Writing in Chemistry: Scientific Papers, Research Grant and Book Proposals

2. **Your contact information:**

   Name: Andrew Terhemen Tyowua
   Affiliation: Benue State University, Makurdi, Nigeria
   Address: Department of Chemistry, Benue State University, Makurdi, Nigeria
   Telephone: +23490918XXXXX
   Email: atyowua@bsum.edu.ng

3. **Will your book be authored or edited?** The book will be authored.

4. **Please provide a one-sentence overview that explains what the book is about, capturing the essence of the book's value and focusing on book content (rather than general subject area).**

   Postgraduates and early career researchers in chemistry struggle with writing scientific papers, research grant and book proposals; therefore, this book is aimed at providing a step-by-step practical guide on how to write scientific papers, research grant and book proposals in chemistry.

5. **Please explain what the reader will gain from adopting or purchasing the book (benefits) and how the book will deliver it (features). Please try to list at least five specific marketable features of your book.**

   Successful completion of postgraduate studies, especially PhD, and career advancement in the academia strongly depend on the ability to publish scientific papers and/or books. Writing research grant proposals is equally important as money must be gotten to conduct research that will lead to a publication (*i.e.*, a research paper or a book). However, many chemical scientists find preparing scientific papers, research grant and book proposals difficult. This is partly because of insufficient training in writing and partly because there are few practical books to enable them to learn the art. By adopting this book, the reader will:

(a) Learn the art of writing scientific papers (research papers, communications and review papers) in chemistry.
(b) Learn the art of writing research grant proposals in chemistry.
(c) Learn the art of preparing book proposals in chemistry.
(d) Improve their chances of getting their papers published in chemistry journals.
(e) Improve their chances of winning research grants in chemistry.
(f) Improve their chances of getting their book proposals accepted.

There are many books on writing scientific papers and grant proposals in the medical sciences, but these are difficult to come by in the chemical sciences. Unlike the few books available on writing in chemical sciences, this book takes a lead by the hand approach and has the following features:

(a) It is straight to the point.
(b) It is written in a simple language to enable easy understanding.
(c) It contains practical examples taken from easily understandable published papers and successful research grant proposals.
(d) It contains chapters on the preparation of graphical abstracts and research highlights, both of which are hardly found in any of the existing books.
(e) It contains chapters on preparation of research grant and book proposals which are rarely found in the existing books.
(f) It contains tasks, taken from easily understandable materials, to help the reader gain hands-on experience.
(g) Sketches/illustrations are used extensively to aid the mental visualization of concepts.

6. **Please describe the type or group of customers for whom the book is intended.**

The book is intended mainly for postgraduate students and early career researchers in chemical science and the libraries that serve them, but can also be useful to other scientists.

7. **Please list below published books one might consider as similar to your own: on the same topic, written at the same level, and intended for the same audience. If you feel there is no direct competition for your book, please list those titles that are more generally related to your book.**

I am not aware of any direct competition for the book; however, there are numerous related titles with the outstanding ones listed in Table AIV.1. Although these books are good and contain useful information on how to write a research paper, I cannot recommend them to chemical scientists because they are biased toward other disciplines and they lack information on other important aspects of writing and scientific genre. For example, many of these titles do not contain information on:

(a) How to write a communication or a review paper?

(b) How to prepare graphical abstracts and research highlights?

(c) How to write research grant and book proposals?

Additionally, these books contain examples that are difficult to understand by readers outside the disciplines of the authors and contain little or no tasks to stimulate hands-on experience. The bottom line is that chemical scientists will gain very little from these books and that a book that addresses these issues (like the one proposed here) is inevitably needed.

Also, please indicate how your book is better or different compared to the competition or related titles. Please be as specific as possible with the differences, as this helps our sales reps a great deal when trying to sell your book.

Unlike the existing related books, the proposed book is targeted at chemical scientists, meaning they will gain more from it compared with those currently on the market, written with bias toward other disciplines. More importantly, the proposed book will contain valuable information on aspects of writing and scientific genre not contained in the ones on the market like:

(a) How to write a communication or a review paper.

(b) How to prepare graphical abstracts and research highlights.

(c) How to write research grant and book proposals.

It also contains numerous examples and various tasks to stimulate hands-on experience.

TABLE AIV.1
Existing Related Books

Title/Author	Publisher	Price	Differences
Mastering Scientific and Medical Writing (A Self-help Guide)	Springer	$864.56	It is biased toward medical science and does not contain: tasks, information on preparation of graphical abstracts and research highlights as well research grant and book proposal writing
Writing Science (How to Write Papers that get Cited and Proposals that get Funded)	Oxford University Press	$155.00	It is biased toward biological science and does not contain: tasks, information on preparation of graphical abstracts and research highlights. There is a quick section on research grant writing, but there is no information on writing book proposals

(Continued)

TABLE AIV.1 (*Continued*)
Existing Related Books

Writing Scientific Research Articles (Strategy and Steps)/ M. Cargil and P. O'Connor	Wiley-Blackwell	$99.44	It is biased toward plant science and does not contain tasks, information on preparation of graphical abstracts and research highlights and grant writing information
The ACS Style Guide (Effective Communication of Scientific Information)/ A.M. Coghill and L.R. Garson (Eds)	Oxford University Press	$67.48	It is meant for ACS journals; does not contain tasks; does not contain information on preparation of graphical abstracts and research highlights; does not address research grant and book proposal writing
How to Write and Illustrate a Scientific Paper/B. Gustavii	Cambridge University Press	$61.05	It is biased toward medical science and does not contain: tasks, information on preparation of graphical abstracts and research highlights and grant writing
From research to manuscript (A Guide to Scientific Writing)/ M.J. Katz	Springer	$59.95	It is biased toward biology and does not contain tasks, information on preparation of graphical abstracts and research highlights and grant writing
Science Research Writing (For Native and Non-native Speakers of English)/ H. Glasman-Deal	World Scientific	$58.00	Does not contain information on the preparation of graphical abstracts and research highlights and grant writing
How to write and publish a scientific paper/ R.A. Day and B. Gastel	Greenwood Press	$49.96	It covers preparation of research papers, but lacks information on preparation of review papers, graphical abstract and research highlights as well as research grant and book proposals
Academic Writing and Publishing (A Practical Handbook)/J. Hartley	Routledge	$39.95	It is biased toward psychology and does not contain tasks, information on preparation of graphical abstracts and research highlights as well as research grant proposals

The prices were retrieved from Amazon (24th October 2020).

8. **Please list 5–6 search terms someone might use in an online search to locate a book on this topic.** Scientific writing (in chemistry); grant writing

(in chemistry); publishing (in chemistry); chemistry writing guide; and grantsmanship (in chemistry)

9. **When will the manuscript be completed and sent to CRC Press?** The manuscript will be completed and sent to CRC Press within 12 months of signing the contract

10. **What is your estimate of the number of pages or word count in the final, double-spaced manuscript?** ~200 papers

11. **How many figures are likely to be included in your manuscript (estimate)?** 30 figures

12. **Is color necessary for any of the figures and how many are likely (estimate or percentage)** Yes, color is necessary for ~35% of figures

13. **How many tables are likely to be included (estimate)?** 12 tables

14. **Please confirm that you will prepare your text in MS Word or please indicate what program you will use to prepare your text?** I confirm that the book will be prepared in MS Word

15. **What program(s) will you use to prepare your camera-ready figures? (*We will supply a list of usable formats, but it is helpful to know upfront what you plan to work with.*)** I aim to use Microsoft PowerPoint to prepare all the figures

16. **How many equations are likely to be included (estimate)?** 15–20 equations will be included in the manuscript

17. **Approximately how much should a book like yours cost?** $40 per copy

18. **In which specific countries may readers have a particular interest in your book, and why?** The proposed book will interest readers globally because the issues that are going to be covered are important to chemical scientists globally

19. **Which societies are applicable for marketing your book?**
    (a) The Chemical Society of Nigeria
    (b) The American Chemical Society
    (c) The Royal Society of Chemistry
    (d) The Chemical Society of Japan

20. Please list any specific magazines or journals for this audience that publish book reviews so that we can be sure to approach them for a review of your book (see Table AIV.2).

TABLE AIV.2

Book Review Journals and Magazines

Nature

Science

Nature Magazine

New Scientist

Philosophical Magazine

---

21. Please list three to five possible reviewers for this proposal, including their names and email addresses. We will contact them to review this proposal and any materials that you supply. If there are any materials, you would like us not to distribute to potential reviewers, please specify.

Name	Email Address

Please attach to this form the following materials:

☑ A proposed table of contents
☑ A brief biography of you and any primary co-authors / editors
☒ Any sample chapters or a preface that may aid in the evaluation of your proposal
☒ A list of people contributing to your book, if any, along with their affiliations

Scan QR code for the attached files and review comments

Index

Note: **Bold** page numbers refer to tables and *italic* page numbers refer to figures.